1000
Sports Cars

1000 Sports Cars

The Most Beautiful and Fastest Cars of All Eras

Author: Reinhard Lintelmann
Producing: SAW Communications, Redaktionsbüro Dr. Sabine A. Werner, Mainz
Translation: Cosima and Nidal Talhouni
Complete production: Naumann & Göbel Verlagsgesellschaft mbH
Printed in China

ISBN 978-3-625-11353-9

Contents

Preface

WAF-AH752

Shining glory from the past

Sports cars combine speed, sporty elegance and the high art of engineering. They are the highlight of every car show, subjects of the longings of old and young car lovers alike. While the names of forgotten car brands like Bizzarrini, Delahaye or Lagonda melt on the tongues of vintage car collectors; the young have also already started to idolise Ferrari, Jaguar and Lamborghini.

However, despite all the admiration and collective passion, the question of "What is a sports car?" remains unanswered. There is no official definition of this term just like there is no unanimous answer to the question of when the first sports car was built. Quite a few chroniclers are of the opinion that sports car history was written already in 1906. On a fine day in June of that year, an automotive sports competition took place for the first time in relation to the Grand Prix on the racetrack of Le Mans, which to this day holds a prominent position in car racing.

In 1914, Fiat introduced the 14B Corsa, a model conceived primarily for competitive car racing. Among other measures to save weight, this model made do without a full chassis and wings.

The Jaguar SS100 is a "Milestone Car" in various aspects. The high speed of 100 mph (approx. 160 km/h) attainable in its cubic capacity class was sensational for its time – reason enough to discreetly integrate this value in the model name.

The Grand Prix competition showed how important it was at that time already for automotive manufacturers to have a presence in car racing. Active participation substantially increased sales figures, while the experience gained from racing could be applied towards the technical advancement of serially produced vehicles. During the time of the First World War, modern technology already allowed many cars to have progressive engines of a type that could barely have been produced ten years earlier. At that point, at the latest, the struggle over performance and speed began in earnest. When in the 1920s regulations limiting the cubic capacity took effect, the designers employed all possible means to extract even the last bit of horsepower from an engine. However, cars with charged compressor engines not only roared along racetracks, "private drivers" who could afford it also enjoyed moving about in sensational sports cars.

Sports cars made by Alfa Romeo, Bentley, Bugatti and Mercedes-Benz were clearly among the automotive highlights of the late 1920s and early 1930s. In addition to these "large" brands, several dozen other manufacturers who developed similarly interesting sports cars also provided car enthusiasts with many topics for conversation. Admittedly, the technology employed by these carmakers did not always correspond to the sportive chassis shape. Nevertheless, cars of lower cubic capacity classes (for instance, BMW, Fiat or MG) sold very well – the most important selling point being their looks.

In 1937, Railton introduced the "Light Sports," an elegantly designed two-seater. A large 4.2-litre, straight eight-cylinder engine powered the impressive car. The silky-soft smooth engine was incidentally a large-scale production engine by Hudson.

Under the bonnet of the Ferrari 250 GT Spyder California the finest technology can be found – the twelve cylinders of the 280 hp engine are arranged in V form. The car which was launched in 1957, was considered by the trade press to be among the most beautiful and most exclusive – but also most expensive – models of its time.

Monteverdi is among the few automotive manufacturers headquartered in Switzerland. In the 1960s and 1970s, Peter Monteverdi, the CEO of the company, was able to achieve a skilful symbiosis of Italian chassis design and American large-scale production technology. The car's engine, a beefy V8 cylinder – was provided by Chevrolet.

While the established and large carmakers continued to develop cubic capacity giants in the 1930s, smaller sports car models became widely popular, especially in Britain. In addition to the noble cars by Aston Martin, Bentley or Jaguar, affordable roadsters such as the MG, Morgan or Triumph gained an increasing foothold on the island. The seat positioning of these models, which was low and often shifted far to the back, provided a driving sensation that is difficult to describe in words. The life of many typically British sports cars often began in the backyard workshop, which did not decrease – and frequently even increased – their sales success and popularity.

Following the Second World War, cars were initially only a means of locomotion and transportation. Instead of high-quality sports cars, production mainly focused on cost-effective small cars and trucks – elegant convertibles and coupes were rarely found until the early 1950s. Then, finally, they arrived – the dream cars everyone had awaited for so long. For most people, however, they remained an illusion, because only a few wealthy individuals were lucky enough to park a gull-winged Mercedes 300 SL in their garage. Nevertheless, everyone else continued to dream – even if only about a Karmann-Ghia. The noble version of the VW bug certainly had a sportive "encasing", yet its road performances left much to be desired. In short: Not every sporty looking car was a "true" sports car, even if its advertisements strongly suggested this.

Similar to England in the late 1930s, some automotive manufacturers considered any open, low two-seater or sleekly styled convertible to be a sports car, regardless of its engine power. Nevertheless, there were some exceptions. Manufacturers like Ferrari, Jaguar or Porsche – to name only a few – had a totally different definition. According to them, a proper sports car could only be produced with a solid foundation. The driving pleasure was determined by engine

The "Museum of Modern Art" in New York does not considers the Lamborghini Miura (year of construction: 1966) to be a mere car in the conventional sense. Rather, the classic car is considered to be an art object on wheels. With some luck, the Miura may be produced again – a prototype featuring the retro design was introduced in 2006.

power, speed and road-holding. Of course, all this came at a price, which is why there continued to be a demand for sports cars in the lower price segment.

While in the 1950s only a limited group of buyers was ready to invest their savings into an expensive sports car, a decade later the situation was quite different. Automotive lovers who had the room and the corresponding financial means began to search for the classics of former days to restore, collect and drive them. Today, however, the scene has changed substantially so that many of these early collections no longer exist. Individuals who buy a classic car today also toy with the idea of it being a potential investment. However, the likelihood of making a quick profit with a vintage car is by no means a certainty.

An interesting concomitant of the passion for collecting is the fact that during the last years an increasing number of historically significant cars have reappeared of whose existence even experts barely had a notion. Many of these finds have frequently changed owners in a very short time. Vendors benefited from the fact that, as a result, the value of many of these rarities more than doubled within a very short period of time and it looks like this boom is likely to continue.

There is therefore no clear answer to whether it is more sensible to invest in a current sports car or in a vintage car. At any rate, serious collectors are primarily interested in technology and design – and not in the potential rise in value. The automotive industry has clearly recognised the market situation.

For many years, the manufacturers of luxury cars have already been constantly increasing the number of produced cars, making sure that the market continues to boom. The supply of classic cars is therefore secured because vintage cars are constantly "replenished" this way.

1905 – 1945

The affection for old cars remains unshaken

Fascination on four wheels

100 years ago, on 27th of June 1906, a Renault's sports car won the world's first Grand Prix. This triumph at the "Grand Prix of France" not only constituted the foundation of Renault's continuously successful racing history. It also made other automotive manufacturers quickly recognise the advantage to their image that could be derived from participating in racing.

Engines and bodies of the Grand Prix vehicles relied for a long time on a heavy, conventionally riveted steel frame. These cars therefore barely differed from the models driven by private persons on the street – except perhaps in terms if cubic capacity and performance. It was not until later that race cars differed in some aspects from the street versions. They profited from numerous technical improvements such as hydraulic shock absorbers or power transmission to the rear axle by a three-gear transmission system as well as cardan shafts. Because of the rather narrow rear track and the fact that tracks mainly consisted of long straight stretches with few curves, the manufacturers often liked to make do without the differential in race cars. This way, some weight could be saved in favour of higher top speeds. The designers were totally free in designing the car engines and as a result a broad range of cubic capacities could be found starting initially at a cubic capacity

of up to 18 litres! It was not until the early 1920s that comparable maximum capacities could be achieved with smaller engines, which did not mean, however, that the cubic capacity giants were discontinued.

Many of the automotive brands that made racing history until the beginning of the Second World War disappeared over the course of time. However, they have left a legacy of many interesting models both for the racetrack or the regular road. Frequently, ambitious private drivers also became the talk of the town with absolutely "normal" street models. These included the industrialist François Repusseau who embarked in 1925 on a journey from Paris to Tunisia in a Renault 40 CV. During his ten-day trip, he drove to his destination via Spain, Morocco and Algeria. Trips of this kind undoubtedly used to be real and rather dangerous adventures.

In the 1930s, an increasing number of sports cars turned out to be true models of reliability. The Monte Carlo rally was in full swing and keenly followed by racing enthusiasts all over Europe. For the general public, the world's most famous street race was a welcome change, helping them forget the problems of those times for a few days. Automotive manufacturers were interested in this race for quite different reasons – "Monte" was one of the most important opportunities to enhance their image. The car make that won the race could justifiable expect to enhance its international prestige coupled with steeply rising sales figures.

In the meantime, it also became apparent that a sports car was not only defined by its, preferably powerful, engine performance but also by its appearance and price. To meet all expectations and reach all groups of buyers, even well-known brands rounded down their model range and constructed, in addition to large six- and eight-cylinder engine cars, also affordable alternatives with only four cylinders. Vehicles of this class sold very well, because the sporty looking chassis constructions of the four-cylinder cars exactly matched the preferences of car enthusiasts. Many of these smaller cars developed into technical masterpieces and constantly revolutionised the young automotive history – never before had technology and design complemented each other this harmoniously.

Audi Type C Alpensieger

When August Horch founded the Audi Werke in 1910, he introduced cars with high cubic capacities like the Type C. From 1912 to 1914, this car model repeatedly won the "Internationale Österreichische Alpenfahrt", which was among the most difficult long distances rallies of those times. Because success had always been an excellent marketing tool, from then on the sporty car was launched on the market under a new name – "Alpensieger" (alpine winner).

Model	Audi Type C Alpensieger
Cubic capacity/cylinder	3564 cc/4 cylinder
hp/kW	35/26
Construction period	1912–1921
Number of produced cars	–

BMW 315/1

At the 1933 Berlin motor show, the stand of BMW featured a prototype of a sports roadster with an especially beautiful shape. Against the rules of the day, the engine was equipped with three instead of two carburettors. Based on the positive feedback, the large-scale production of the car began in 1934 – the gem could be purchased for 5200 Reichsmark.

Model	BMW 315/1
Cubic capacity/cylinder	1490 cc/6 cylinder
hp/kW	40/29
Construction period	1934–1935
Number of produced cars	230

BMW 319/1

Many owners of a BMW 315/1 Roadster soon noticed that they could be very successfully involved in racing with this model. The manufacturer reacted immediately and developed an almost identical car in the two-litre class. Basically, they only drilled the engine open – resulting in barely any serious optical differences between the 315/1 and 319/1 models.

Model	BMW 319/1
Cubic capacity/cylinder	1911 cc/6 cylinder
hp/kW	55/40
Construction period	1934–1936
Number of produced cars	178

BMW 328

In the mid-1930s, BMW very quietly developed this sports car which soon received great fame and ensured BMW one of the foremost positions in international racing history. Even though BMW was already among the most famous automotive manufacturers, the competition introduced increasingly stronger models, which made it imperative for BMW to come up with its own top model.

Model	BMW 328
Cubic capacity/cylinder	1971 cc/6 cylinder
hp/kW	80/59
Construction period	1936–1939
Number of produced cars	464

BMW 328 Coupé

In 1936, the BMW 328 was built exclusively for racing. Private drivers who wanted to buy a 328 could only purchase the model as of 1937. By the way: two special racing models featuring and extra-light chassis and tuned engines gained a class victory in Le Mans. As an ultimate highlight, chassis builder Wendler in Reutlingen also created two streamlined coupés.

Model	BMW 328
Cubic capacity/cylinder	1971 cc/6 cylinder
hp/kW	135/99
Construction period	1936 – 1939
Number of produced cars	2 coupés

BMW 327

To offer private drivers an affordable sports car, BMW developed in 1936 another 2+2 seat sports convertible. The so-called BMW 327 was partially based on the 328, but was only equipped with a 55 hp engine. For a surcharge, an 80 hp engine was also made available later on. Vehicles with this configuration were offered on the market under the BMW code 327/28.

Model	BMW 327
Cubic capacity/cylinder	1971 cc/6 cylinder
hp/kW	55/40
Construction period	1937 – 1941
Number of produced cars	1306

Horch 670

The automotive highlights exhibited by manufacturers at the 1931 Paris Salon included an elegant sports convertible by Horch. Horch had always been well represented in the luxury class. However, what it presented here even took away the breath of the professional world: The new Horch 670 (the competitive model to the Maybach Zeppelin) was powered by a twelve-cylinder engine.

Model	Horch 670
Cubic capacity/cylinder	6021 cc/12 cylinder
hp/kW	120/88
Construction period	1931 – 1934
Number of produced cars	approx. 80

Horch 853 A

In 1935, Horch presented the so-called "Horch 5 litre" series for the first time. This series was absolutely topped by the prestigious 853 A model, which was primarily styled as a quick sports convertible. Cars of this class were an absolute must for the high society of the time. If desired, they could even have the 853 A fitted with a special customised coachwork – everything was possible.

Model	Horch 853 A
Cubic capacity/cylinder	4944 cc/8 cylinder
hp/kW	120/88
Construction period	1938 – 1939
Number of produced cars	approx. 1000

Mercedes Simplex 28/32 hp

While the majority of cars around the year 1900 still largely resembled coaches, the design of the early Mercedes models already generated a great amount of attention. They were based on a solid chassis and were equipped with powerful engines, the best prerequisites for the young sport of car racing. Their performance could be admired during the racing week of Nice and many similar events.

Model	Mercedes Simplex 28/32 hp
Cubic capacity/cylinder	5315 cc/4 cylinder
hp/kW	32/23
Construction period	1901 – 1905
Number of produced cars	–

Mercedes Grand Prix Race Car

The first car racing event was established in 1894. It consisted of a car reliability drive from Paris to Rouen. Participation in the drive was a matter of course for the Daimler-Motoren-Gesellschaft. Many of the insights gained in this event were later used in the development of the first Grand Prix race car.

Model	Mercedes Grand Prix Race Car
Cubic capacity/cylinder	7280 cc/6 cylinder
hp/kW	90/66
Construction period	1914
Number of produced cars	–

Mercedes Simplex 60 hp

Because the Daimler-Motoren-Gesellschaft only produced high-calibre vehicles and as a matter of principle always opted for high-quality technology from the beginning, the Mercedes brand was always one step ahead of its competition. For example, as of 1902, each and every one of their models already included and an ingenious device for the control and dosage of oil, the so-called drip-feed lubricator.

Model	Mercedes Simplex 60 hp
Cubic capacity/cylinder	9235 cc/4 cylinder
hp/kW	60/44
Construction period	1902 – 1905
Number of produced cars	–

Mercedes 24/100/140 hp

Headed by chief construction engineer Dr. Ferdinand Porsche, a new era was launched at the Daimler-Motoren-Gesellschaft in 1921. Porsche perfected the compressor technology and developed the 24/100/140 hp, an especially sporty car. Under the normal driving conditions, its six-cylinder engine had a performance of 100 hp, if the compressor was additionally switched on, the potential performance even increased to 140 hp.

Model	Mercedes 24/100/140 hp
Cubic capacity/cylinder	6240 cc/6 cylinder
hp/kW	100/73
Construction period	1924 – 1925
Number of produced cars	–

Mercedes-Benz K 24/110/160 hp

The prototype "K" was developed in 1926 as a further development of the Mercedes 24/100/140 hp. This sporty vehicle remained in the program under the model name Mercedes-Benz K 24/110/160 hp following the merger of the Daimler and Benz companies. Since the "K" was based on a solid chassis with relatively short wheel base, it was relatively easy to implement different coach constructions.

Model	Mercedes-Benz K 24/110/160 hp
Cubic capacity/cylinder	6240 cc/6 cylinder
hp/kW	110/81
Construction period	1928 – 1929
Number of produced cars	

Mercedes-Benz S 26/120/180 hp

The Mercedes-Benz S could not have had a better start – it celebrated its launch at the opening race of the Nürburgring, crowning the sporty event at the same time with a double victory! In technical terms, the "S" is an improved development of the 630 K model. Without the compressor the engine transmitted 120 hp to the rear axle, with the compressor, this value was increased to 180 hp.

Model	Mercedes-Benz S 26/120/180 hp
Cubic capacity/cylinder	6800 cc/6 cylinder
hp/kW	120/88
Construction period	1926 – 1930
Number of produced cars	174

Mercedes-Benz SS

When Daimler-Benz juxtaposed the model SS to the Mercedes-Benz Type S in 1928, it was decided from the start that a certain number of SS chassis frames would be fitted with special bodies. Thus customers were able to add a totally personal touch to their SS. The SS shown here, for example, was equipped with coachwork created by the Italian firm of Castagna.

Model	Mercedes-Benz SS
Cubic capacity/cylinder	7065 cc/6 cylinder
hp/kW	170 (with compressor 225)/125 or 165
Construction period	1928 – 1934
Number of produced cars	115

Mercedes-Benz SS 27/140/200 hp

While the majority of Mercedes-Benz models S, SS, SSK and SSKL were equipped with open racing bodywork; the product range also included saloons and convertibles. In the second development stage, called SS (= Super Sport) the vehicles were even more powerful due to a more compressed engine and dual ignition.

Model	Mercedes-Benz SS 27/140/200 hp
Cubic capacity/cylinder	7065 cc/6 cylinder
hp/kW	140/103
Construction period	1928 – 1932
Number of produced cars	151

Mercedes-Benz SSKL

The Mercedes-Benz SSKL, offered in the market at a price of 40.000 Reichsmark, was nothing other than a pure racing tool, which had been subject to various slimming measures in comparison to the SSK. To keep its weight as low as possible, measures included boring the chassis. The SSKL (Super Sport Kurz Leicht (= short light)) was primarily used for racing and sold exclusively to racing drivers.

Model	Mercedes-Benz SSKL
Cubic capacity/cylinder	7065 cc/6 cylinder
hp/kW	300/220
Construction period	1928 – 1933
Number of produced cars	7

Mercedes-Benz SSK

Without doubt, a Mercedes-Benz SSK (Super Sport Kurz (= short)) was the absolute dream car of its times for drivers with racing ambitions. The SSK – principally still based on the basic model Type S (Sport) – featured a particularly short wheel base. This turned it into the ultimate racing equipment. Rudolf Caracciola and other racing drivers regularly achieved victories driving the SSK.

Model	Mercedes-Benz SSK
Cubic capacity/cylinder	7065 cc/6 cylinder
hp/kW	140/103
Construction period	1928 – 1932
Number of produced cars	42

Mercedes-Benz 380

In 1933, Daimler-Benz introduced the Type 380 to the market, a vehicle whose most important construction feature was the use of so-called independent axles, since the rigid axles that had been in use up to that time were discontinued. The technology of the engine was also improved and the 380 was equipped with an overdrive transmission, which could lower the engine revolutions on highway journeys.

Model	Mercedes-Benz 380
Cubic capacity/cylinder	4019 cc/8 cylinder
hp/kW	90 (with compressor 144)/66 or 105
Construction period	1933 – 1934
Number of produced cars	154

Mercedes-Benz Mannheim 370 S

The Mercedes-Benz model 370 S – better known under the model name "Mannheim" – is an example of the fact that it is possible to create a sporty two-seater from the technical components of a limousine. The owners of the 370 S included the racing driver Rudolf Caracciola – who drove his Mannheim 370 S as a second car.

Model	Mercedes-Benz Mannheim 370 S
Cubic capacity/cylinder	3689 cc/6 cylinder
hp/kW	75/55
Construction period	1930 – 1933
Number of produced cars	183

Mercedes-Benz 500 K

While in the late 1920s the Mercedes-Benz models S, SS and SSK received many headlines in the automotive press, as of 1934 the abbreviation that caused the most euphoria was "500 K". The sporty 500 K was just the ticket for affluent customers – it offered them elegance and comfort while surprising them with impressive performance. On the technical side, it pampered its owners by offering the first-ever single-wheel suspension.

Model	Mercedes-Benz 500 K
Cubic capacity/cylinder	5018 cc/8 cylinder
hp/kW	100 (with compressor 160)/73 or 117
Construction period	1934 – 1936
Number of produced cars	342

Mercedes-Benz 500 K

In 1934, with the model 500 K, Daimler-Benz once again offered a vehicle of the ultimate deluxe class. Under its long bonnet it had an eight-cylinder compressor engine which had an output of 160 hp when the compressor was switched on. The 500 K was offered in all bodywork variations, yet it was basically designed to be a rather elegant two- to four-seater sports car.

Model	Mercedes-Benz 500 K
Cubic capacity/cylinder	5018 cc/8 cylinder
hp/kW	100 (with compressor 160)/73 or 117
Construction period	1934 – 1936
Number of produced cars	342

Mercedes-Benz 500 K Spezial Roadster

The 500 K Spezial Roadster was a design masterpiece, stylishly elegant as well as shockingly expensive. For this two-seat dream on wheels, Daimler-Benz used a chassis frame with a shortened wheel base, because this was the only way to optimally proportion the roadster's carriage. As opposed to the "normal" 500 K, the massive radiator was positioned not in front of the front axle, but directly on top of it.

Model	Mercedes-Benz 500 K Spezial Roadster
Cubic capacity/cylinder	5018 cc/8 cylinder
hp/kW	100 (with compressor 160)/73 or 117
Construction period	1934 – 1936
Number of produced cars	38 on short chassis

Mercedes-Benz 540 K

Owners of the 500 K had actually no reason whatsoever for complaint. They owned a car of the deluxe class which only few people could afford. However, the 500 K by no means represented the ultimate in car luxury. In 1936, the car maker launched the even more powerful 540 K. It easily reached speeds of 170 km/h while consuming about 27 to 30 litres of gasoline per 100 kilometres.

Model	Mercedes-Benz 540 K
Cubic capacity/cylinder	5401 cc/8 cylinder
hp/kW	115 (with compressor 180)/84 or 132
Construction period	1936 – 1939
Number of produced cars	406

Mercedes-Benz 540 K

The main distinguishing feature of the 540 K was primarily the fact that it allowed relaxed travelling with consistently high speed. However, on winding roads the car weighing 2300 kg had a rather hard time. In constructing the engine, the engineers had focused primarily on extreme quietness. The straight eight-cylinder engine was therefore more than one metre long and consisted of sound-absorbing special grey iron.

Model	Mercedes-Benz 540 K
Cubic capacity/cylinder	5401 cc/8 cylinder
hp/kW	115 (with compressor 180)/84 or 132
Construction period	1936 – 1939
Number of produced cars	406

Mercedes-Benz 540 K

As common for deluxe cars of that era, the price of the vehicle consisted of the cost of the bare chassis plus the bodywork construction. The bodywork for the 540 K cost on average about 22.000 Reichsmark – more than the chassis. In addition, Daimler-Benz also complied with special customer requests (special equipment, paint work, etc.) which could increase the price even further.

Model	Mercedes-Benz 540 K
Cubic capacity/cylinder	5401 cc/8 cylinder
hp/kW	115 (with compressor 180)/84 or 132
Construction period	1936 – 1939
Number of produced cars	406

Opel Rennwagen

In 1913 it there was hardly a more sporty vehicle to be found: Opel set a 110 hp two-seater on rather high wheels and equipped the four-meter long model with a straight block construction engine. Despite its massive weight and the use of heavy materials (grey iron for the engine block) the racer was able to achieve 170 km/h!

Model	Opel Rennwagen
Cubic capacity/cylinder	3970 cc/4 cylinder
hp/kW	110/81
Construction period	1913
Number of produced cars	–

Wanderer W 25 K

In 1936, Wanderer, a Auto Union AG brand, introduced an interesting sports car intended to compete with the BMW 328 – the Type W 25 K. To make it especially agile, its engine was coupled with a constantly operating compressor which, increased its output. Unfortunately, this demanded of the engine more than it could handle, resulting in frequent warranty claims.

Model	Wanderer W 25 K
Cubic capacity/cylinder	1950 cc/6 cylinder
hp/kW	85/62
Construction period	1936 – 1939
Number of produced cars	258

Minerva AKS

As expected, Minerva (Belgium) in the early 1930s followed the model policy of its competitors and introduced especially powerful cars to the market. This AKS model, also known as "Speed Six," was exactly in line with the preferences of the time and distinguished itself from its predecessors by numerous improvements (sevenfold mounted crankshaft, optimised transmission ratio, etc.).

Model	Minerva AKS
Cubic capacity/cylinder	5954 cc/6 cylinder
hp/kW	150/110
Construction period	1930 – 1931
Number of produced cars	–

Alvis 12/75 F.W.D.

In 1919, along with several others, car maker Alvis settled in Coventry, the stronghold of the British automotive industry at that time. Alvis aimed to only introduce to the market high-quality, and above all sporty-looking, vehicles. Very courageously, the company already tried in 1928 to launch a car with the still relatively unknown front-wheel drive.

Model	Alvis 12/75 F.W.D.
Cubic capacity/cylinder	1482 cc/4 cylinder
hp/kW	50/37
Construction period	1928 – 1929
Number of produced cars	–

Alvis Silver Eagle

1928 was a rather important year for the British brand Alvis since it was the year it presented its model F.W.D., the world's first six-cylinder car with front-wheel drive. To also meet the needs of less demanding customers, the company additionally developed some models with a rather more moderate engine power. A good decision – the general public had just been waiting for an Alvis model such as the "Silver Eagle".

Model	Alvis Silver Eagle
Cubic capacity/cylinder	2148 cc/6 cylinder
hp/kW	72/53
Construction period	1928 – 1934
Number of produced cars	–

Alvis 12/60 hp

After presenting a front-wheel drive sports car, Alvis brought out several four-cylinder models with conventional rear-wheel drive. To vividly demonstrate the speed and easy handling of the model 12/60 hp, the company selected the figure of a hare as its hood ornament – which perfectly symbolised all the distinguishing properties of this car.

Model	Alvis 12/60 hp
Cubic capacity/cylinder	1645 cc/4 cylinder
hp/kW	50/36.6
Construction period	1931 – 1932
Number of produced cars	–

Alvis Speed Twenty

The Alvis Speed Twenty, a vehicle primarily developed for car racing, was supposedly constructed within 14 weeks only. It was based on a very modern and light chassis. In addition, the Twenty featured central chassis lubrication and – of particular importance for use in racing – snubbers with adjustable hardness.

Model	Alvis Speed Tweety
Cubic capacity/cylinder	2511 cc/6 cylinder
hp/kW	87/64
Construction period	1932 – 1934
Number of produced cars	–

Aston Martin 1.5 Litre

In 1908 Lionel Martin and Richard Bamford had the idea of putting together a "true" sports car. In the beginning, they used Isotta-Fraschini chassis for their experiments, before venturing into independence in 1922. In that year they introduced to the market under the trade name Aston Martin their first four-cylinder vehicles with chassis manufactured by themselves.

Model	Aston Martin 1.5 Litre
Cubic capacity/cylinder	1493 cc/4 cylinder
hp/kW	60/44
Construction period	1934 – 1936
Number of produced cars	–

Bentley 6½ Litre

While the 1920s saw many cars with gorgeously distinguished bodywork, their sporty appearance could not, however, conceal that fact that these vehicles often lacked in agility – they were simply too heavy. This is why Bentley developed an especially powerful 6.5-litre engine with which it equipped its larger vehicles with long wheel bases.

Model	Bentley 6½ Litre
Cubic capacity/cylinder	6597 cc/6 cylinder
hp/kW	145/106
Construction period	1926 – 1930
Number of produced cars	363

Bentley 4½ Litre Blower

Although the Bentley models of the 4½-litre class achieved many victories, to deriding critics these cars were nothing more than "quick trucks". W.O. Bentley could handle the criticism and enjoyed demonstrating the potential of his cars. With the help of the compressor technology, as it was used in the "Blower Bentley", he extracted even more power from the engines.

Model	Bentley 4½ Litre Blower
Cubic capacity/cylinder	4398 cc/4 cylinder
hp/kW	182/133
Construction period	1927 – 1931
Number of produced cars	55

Bentley 4½ Litre

Walter Owen Bentley had only one goal: In 1919 he wanted to build the first typically British sports car. Initially, he experimented with four-cylinder engines with three-litre cubic capacity and from 1926 he foresaw the future rather in models of the 4½-litre class. In financial terms, Bentley's enthusiastic engagement was a disaster – reason enough to sell the troubled brand to Rolls Royce in 1931.

Model	Bentley 4½ Litre
Cubic capacity/cylinder	4398 cc/4 cylinder
hp/kW	110/81
Construction period	1926 – 1930
Number of produced cars	665

Bentley 4 1/2 Litre

Under the slogan "The Silent Sports Car", Bentley introduced some very innovative models to the market as of 1936. For example, the function of the shock absorbers could be adjusted while driving by means of a pump connected to the transmission. By the way, a heavy Bentley required about 20 seconds for the spurt from 0 to 100 km/h – a good value for that time.

Model	Bentley 4 1/2 Litre
Cubic capacity/cylinder	4257 cc/6 cylinder
hp/kW	125/92
Construction period	1936 – 1940
Number of produced cars	–

Frazer Nash TT

Starting in 1924, the Briton Archie Frazer-Nash built sporty two-seaters which he equipped with a "chain drive", as a novelty. He did not manufacture engines himself, but rather relied on the engines of famous manufacturers of the time. It was important to him that these could be excellently tuned. After the Second World War, Frazer-Nash specialised in importing BMW cars.

Model	Frazer Nash TT
Cubic capacity/cylinder	1496 cc/4 cylinder
hp/kW	62/45
Construction period	1937
Number of produced cars	–

(Jaguar) SS 1 Coupé 16 hp

The history of Jaguar began in 1922 when William Lyons established the Swallow Sidecar Company, which focused on the construction of motorcycle sidecars. When he moved in 1928 to Coventry, he prepared for automotive manufacturing. As early as 1931, the Sports Car SS 1 emerged from the factory halls. As part of progressive model development, the engine power was soon increased, adding top models to the series.

Model	SS 1 Coupé 16 hp
Cubic capacity/cylinder	2054 cc/6 cylinder
hp/kW	48/35
Construction period	1931 – 1936
Number of produced cars	4230

(Jaguar) SS 1 Airline 20 hp

To improve the optical appearance of his cars, Lyons developed for the models of the year 1932 an especially low chassis, which was meant to lower the waistlines of the cars. In addition, the new chassis types were especially suited for mounting magnificent bodies – as vividly demonstrated by this Airline coupé of 1936.

Model	SS 1 Airline 20 hp
Cubic capacity/cylinder	2552 cc/6 cylinder
hp/kW	62/45
Construction period	1933 – 1936
Number of produced cars	573

(Jaguar) SS 2–12 hp

Parallel to the six-cylinder SS models, Lyons also built various four-cylinder cars. The basic model of this series was the small SS 2 (9 hp) that only existed as a coupé until 1933. As part of the model development, the performance was constantly increased, and later the vehicle advanced into the model SS 2–12 hp. In addition, all models manufactured as of 1932 profited from the use of a new chassis.

Model	SS 2–12 hp
Cubic capacity/cylinder	1608 cc/4 cylinder
hp/kW	38/28
Construction period	1933 – 1936
Number of produced cars	1800

Jaguar SS 100

In 1935, William Lyons introduced the trade name Jaguar. The first model to carry this name was the two-seater sports car Jaguar SS 100. It achieved a high speed of 160 km/h (corresponding to 100 miles), which was remarkable at the time. The first series of the SS 100 was equipped with a 2.6-litre engine, while the 3.5-litre engine was available as of 1938.

Model	Jaguar SS 100
Cubic capacity/cylinder	2663 cc/6 cylinder
hp/kW	102/75
Construction period	1936 – 1939
Number of produced cars	310

Lagonda 2 Litre

Wilbur Gunn, an opera singer (!), established the Lagonda brand. The native American initially built cycle cars in England, before he advanced to automotive manufacturing in 1906. The brand became famous as of the mid-1920s with advanced constructions like the 2 Litre. This tourer, which alternately also came with a compressor engine, was based on a very low chassis.

Model	Lagonda 2 Litre
Cubic capacity/cylinder	1954 cc/4 cylinder
hp/kW	85/63
Construction period	1925 – 1933
Number of produced cars	–

Lagonda 2 Litre Speed

If only due to its long wheel base of 305 centimetres, the Lagonda 2 Litre Speed model is one of those impressive cars that one enjoys looking at. With a top speed of 130 km/h, the car was not only driven on the road but also in races (Le Mans). The installation of two carburettors allowed all vehicles built from 1930 onward to achieve better acceleration rates.

Model	Lagonda 2 Litre Speed
Cubic capacity/cylinder	1954 cc/4 cylinder
hp/kW	70/51
Construction period	1927 – 1932
Number of produced cars	–

Lagonda M 45

Buyers who were looking for something special found it in the M 45 offered by Lagonda as of 1933. The extremely elegant car held its ground not only on the street but also on the racecourse. It gained first racing victories in 1934 at the Tourist Trophy. One year later, an M 45 travelling at an average speed of roughly 124 km/h won the 24 hours of Le Mans.

Model	Lagonda M 45
Cubic capacity/cylinder	4467 cc/6 cylinder
hp/kW	140/102.5
Construction period	1933 – 1936
Number of produced cars	–

Lagonda Rapid V 12

In 1936, Lagonda surprised the automotive world with a V 12-cylinder car that was based on a completely new and innovative chassis. This base consisted of a x-braced frame with additional reinforcement in the rear area. While the rear wheels were suspended by conventional semi-elliptic springs, the carriage received a very modern torsion-bar suspension in front.

Model	Lagonda Rapid V 12
Cubic capacity/cylinder	4480 cc/12 cylinder
hp/kW	175/129
Construction period	1938 – 1939
Number of produced cars	189

Lagonda LG 6 Rapid

Lagonda's show piece, the large Rapid V 12, was based on a very modern chassis construction which could also be used in a slightly modified form for smaller models – the LG Type was among the models that benefited from this base. The LG 6 was equipped with a six-cylinder engine and offered an alternative to customers who did not want to invest their money in the Rapid V 12.

Model	Lagonda LG 6 Rapid
Cubic capacity/cylinder	4467 cc/6 cylinder
hp/kW	140/102.5
Construction period	1936 – 1939
Number of produced cars	–

Lagonda Rapier 10

It was not unusual for a manufacturer of sports cars to also offer models in the lower cubic capacity classes. Therefore Lagonda also offered a speedy model for beginners, the Rapier. Of high-quality and solid construction like every Lagonda, the Rapier cost about GBP 375. In terms of size, it resembled the MG or Singer, while offering a higher prestige value.

Model	Lagonda Rapier 10
Cubic capacity/cylinder	1086 cc/4 cylinder
hp/kW	55/40
Construction period	1934 – 1939
Number of produced cars	470

MG 14/40 hp

Cecil Kimber, owner of a Morris representation, was skilled in tuning unspectacular Morris cars to make them faster. In 1923, he was able to launch his first own car called MG. He chose the name of MG in reference to his workshop "Morris Garage". It was from precisely this garage that the first MG serial models emerged soon afterwards.

Model	MG 14/40 hp
Cubic capacity/cylinder	1802 cc/4 cylinder
hp/kW	40/30
Construction period	1924 – 1929
Number of produced cars	–

MG PB Midget

When Singer launched the sports car Type "Le Mans", MG had long since been ready with its reply. It was launched in 1935 in the shape of the "PB" that had a higher cubic capacity than its predecessor model PA. Additionally, it also featured an elaborately equipped fascia, as well as a stone shield for the radiator. In 1935, MG made the decision to step up the production of everyday cars.

Model	MG PB Midget
Cubic capacity/cylinder	939 cc/4 cylinder
hp/kW	35/26
Construction period	1934 – 1936
Number of produced cars	approx. 530

MG TA Midget

Because MG models were equipped with many large-production modules, the company could offer its sports cars at especially attractive prices. They consciously made do without an in-house bodywork department, instead commissioning the firm Carbodies with the construction of bodies at 6 pounds each. Luxury was not really desired as, after all, an MG car was intended to remain a financially controllable pleasure.

Model	MG TA Midget
Cubic capacity/cylinder	1292 cc/4 cylinder
hp/kW	52/38
Construction period	1936 – 1939
Number of produced cars	approx. 3000

Morgan Sports

Because H.S.F. Morgan considered two wheels to be too unsafe, he designed a tricycle with which he laid the foundation stone of a company that builds sports cars to this day. Morgan's concept was simple and brilliant – the vehicle was based on a frame consisting of three tubes. A V2-cylinder moved the construction, and depending on the power of the engine, it was even possible to win races with Morgan's "Threewheelers".

Model	Morgan Sports
Cubic capacity/cylinder	990 cc/2 cylinder
hp/kW	32/23
Construction period	1931 – 1934
Number of produced cars	–

Morgan Super Sports

Morgan's tricycles, which initially were not taken very seriously, quickly developed into sporty companions due to the use of high-horsepower engines which helped them repeatedly win races. In the late 1920s, the Threewheeler almost already had the characteristics of "proper" cars: They featured a three-gear transmission as well as perfect brakes and, above all, they sold very well.

Model	Morgan Super Sports
Cubic capacity/cylinder	1096 cc/2 cylinder
hp/kW	40/29
Construction period	1934 – 1937
Number of produced cars	–

Morgan 4/4

In 1936, the Britons lost their interest in purchasing tricycle vehicles. The government cancelled the tax relief on them, and Morgan had to find an alternative as soon as possible. Based on the modified base of the last Threewheeler, a full car with four wheels was developed – the Morgan 4/4. Incidentally, the underside of early models was panelled with floorboards.

Model	Morgan 4/4
Cubic capacity/cylinder	1098 cc/4 cylinder
hp/kW	34/25
Construction period	1936 – 1939
Number of produced cars	–

Morgan 4/4

In 1936, when the first four-wheel car emerged from the factory halls of Morgan, nobody could foresee that this was the launch of a car that would for decades remain almost unchanged in the company's portfolio. The name 4/4 incidentally, indicated that the successor of the Threewheeler now moved on four wheels and was equipped with a four-cylinder engine.

Model	Morgan 4/4
Cubic capacity/cylinder	1267 cc/4 cylinder
hp/kW	40/29
Construction period	1939 – 1957
Number of produced cars	–

Morris 10/6

Following the launch of the small "Eight", Morris had actually been prepared for an annual production of 35.000 units. This was a miscalculation, since 50.000 orders were already placed in the first year. In addition, customers asked for a sportier version, to which Morris reacted promptly and developed the pretty 10/6 based on the "Eight".

Model	Morris 10/6
Cubic capacity/cylinder	1378 cc/6 cylinder
hp/kW	38/28
Construction period	1934 – 1935
Number of produced cars	–

Singer 9 hp Le Mans

Among the sporty cars in the cubic capacity class of up to one litre, the Singer Le Mans enjoyed a particularly good reputation. It was practically a challenge to every tuner to utilise to the fullest the potential inherent in its small machine. Singer charged a modest GBP 225 for the little car – even an MG could not be had at such a competitive price.

Model	Singer 9 hp Le Mans
Cubic capacity/cylinder	972 cc/4 cylinder
hp/kW	39/29
Construction period	1935 – 1937
Number of produced cars	–

Bugatti T 23 Brescia

At the beginning of his career, in addition to his models with high cubic capacity, Ettore Bugatti also produced cars in the 1.5-litre class such as the T 23, which was an especially modern vehicle. The motor was fitted with a 16-valve engine, coupled with a narrowly gradated four gear transmission, while the steering system reacted spontaneously to the smallest of moves.

Model	Bugatti T 23 Brescia
Cubic capacity/cylinder	1496 cc/4 cylinder
hp/kW	30/22
Construction period	1921 – 1926
Number of produced cars	–

Bugatti Type 35 A

Bugatti's legendary Grand Prix car, the Type 35, was based on a chassis that tapered towards the back. This provided the vehicle with its striking rear end section, the so-called "boat stern". The front always featured a horseshoe-shaped radiator whose shape varied slightly from one model to another. For private drivers, Bugatti offered the Type 35 A as an alternative.

Model	Bugatti Type 35 A
Cubic capacity/cylinder	1991 cc/8 cylinder
hp/kW	75/55
Construction period	1926 – 1930
Number of produced cars	130

Bugatti Type 40

In 1926, Bugatti developed a new chassis that was used for the first time as a base for the Type 40. The effort had been worthwhile, since never before did a Bugatti have such pleasant handling characteristics. Due to its construction style, this chassis was not only compatible with convertible or roadster bodies, but also with heavy, closed saloon styles.

Model	Bugatti Type 40
Cubic capacity/cylinder	1496 cc/4 cylinder
hp/kW	45/33
Construction period	1926 – 1930
Number of produced cars	approx. 800

Bugatti Type 57

With the development of the Type 57, a new era began at Bugatti, since Ettore's son Jean had contributed largely to the construction of this new model. The result of this collaboration was well received by the trade press: it considered the Type 57 with its civilised eight-cylinder engine to be the most functional Bugatti of all times.

Model	Bugatti Type 57
Cubic capacity/cylinder	3257 cc/8 cylinder
hp/kW	135/99
Construction period	1934 – 1940
Number of produced cars	approx. 700 (whole series)

Bugatti Type 57 Atalante

While many Bugatti customers fancied customised special bodywork, others opted for a standard serial construction of the firm. Jean Bugatti always designed these body styles. However, the bodies were not produced in-house but by famous coach makers (for instance, Gangloff in Colmar).

Model	Bugatti Type 57 Atalante
Cubic capacity/cylinder	3257 cc/8 cylinder
hp/kW	160/117
Construction period	1937 – 1940
Number of produced cars	approx. 700 (whole series)

De Dion Bouton 8 hp

Even if the automobile was invented in Germany, the French made significant contributions to the development of motorised transportation. For example, the founder of the first French automotive club, Count de Dion, developed improvements such as the V8 engine and a gear that could be switched without grinding. In 1913 the Count ceased to build cars – shortly after launching the model 8 hp.

Model	De Dion Bouton 8 hp
Cubic capacity/cylinder	3122 cc/4 cylinder
hp/kW	15/11
Construction period	1911 – 1914
Number of produced cars	–

Delahaye 135 MS

To prove that the Delahaye 135 MS could reach a high speed of 148 km/h, the company raced this model in 1939 on the Brooklands racetrack against a number of competitors. Delahaye proved to be correct: the Type 135 MS reached its destination ahead of the Alfa Romeo 2.9 litres, the Talbot Lago 4 litres, the Delage 3 litres and other competitors.

Model	Delahaye 135 MS
Cubic capacity/cylinder	3557 cc/6 cylinder
hp/kW	130/95
Construction period	1938 – 1952
Number of produced cars	–

Peugeot Darl'Mat Sport

In 1936, the Paris Peugeot dealer Emile Darl'Mat launched a sports car built according to his visions. For the foundation, he used the chassis of the Peugeot 302 for its wheel base of 288 centimetres. The Darl'Mat was powered by the approximately 160 km/h tuned four-gear engine of the Peugeot 402.

Model	Peugeot Darl'Mat Sport
Cubic capacity/cylinder	1991 cc/4 cylinder
hp/kW	87/64
Construction period	1936 – 1938
Number of produced cars	3

Peugeot 402 Eclipse

With the Eclipse, Peugeot introduced an avant-garde vehicle whose roof design was particularly striking. The foldable metal sheet roof of this coupé convertible could be completely lowered (electrically, if desired) into the boot – an ingenious technology which transformed the vehicle very quickly into a completely open car.

Model	Peugeot 402 Eclipse
Cubic capacity/cylinder	1991 cc/4 cylinder
hp/kW	55/40.3
Construction period	1937 – 1939
Number of produced cars	–

Renault Type K

Around the year 1900, participation in races was already inalienable for automobile manufacturers, because this was the only way for them to impressively demonstrate the quality of their models. From the outset, Renault was among the most active participants in races in which they regularly entered sporty versions, like the Type K.

Model	Renault Type K
Cubic capacity/cylinder	7500 cc/4 cylinder
hp/kW	42/31
Construction period	1902 – 1907
Number of produced cars	–

Renault AK 90 CV

In 1906 Renault won the world's first Grand Prix. The winning car, an AK 90 CV with 13 litres (!) of cubic capacity, reached a high speed of 154 km/h, which was inconceivable at that time. A heavy steel frame supported the engine and the body of the Grand Prix car. Incidentally, the power transmission to the rear axle by a three-gear transmission as well as the cardan shaft were technologically far ahead of their time.

Model	Renault AK 90 CV
Cubic capacity/cylinder	12.995 cc/6 cylinder
hp/kW	90/66
Construction period	1906
Number of produced cars	–

Renault 40 CV

The 40 CV of Renault was among the most interesting cars with giant cubic capacities that received great attention in the 1920s. The almost six-meter-long carriage was produced with numerous bodywork variations which even held their ground in races. The radiator of the 9112 cubic centimetre six-cylinder engine incidentally required 50 litres of coolant!

Model	Renault 40 CV
Cubic capacity/cylinder	9112 cc/6 cylinder
hp/kW	104/76
Construction period	1921 – 1929
Number of produced cars	

Alfa Romeo RM Sport

The Alfa Romeo RL introduced in 1922 was not only the first model of this brand equipped with a six-cylinder engine, but above all a car which experienced race drivers enjoyed. Already one year later, the RM shown here was developed based on the RL. It was matched to the needs of private drivers and equipped with a four-cylinder engine.

Model	Alfa Romeo RM Sport
Cubic capacity/cylinder	1944 cc/4 cylinder
hp/kW	40/30
Construction period	1923 – 1925
Number of produced cars	–

Alfa Romeo 6 C 1750 Sport

For the 1929 model year, Vittorio Jano, a former Fiat engineer who later switched to Alfa Romeo, developed a highly competitive compressor vehicle, the Alfa Romeo 6 C 1750. The 6 C was almost predestined for racing. Drivers like Campari, Nuvolari and Varzi repeatedly scored victories driving it. If requested, the 6 C was also sold to private drivers without a compressor.

Model	Alfa Romeo 6 C 1750 Sport
Cubic capacity/cylinder	1752 cc/6 cylinder
hp/kW	55/40
Construction period	1929 – 1933
Number of produced cars	approx. 320 (whole series)

Alfa Romeo 6 C 1750

Since its establishment in 1931, "S.A. Carrozzeria Pinin Farina", established by Battista Farina (a.k.a. "Pinin") has been co-operating with Alfa Romeo. The first model who was built within the scope of this co-operation was the Alfa Romeo 6 C 1750. This convertible marked the beginning of a high-class co-operation, which resulted in the launch of the most legendary cars of the 20th century.

Model	Alfa Romeo 6 C 1750
Cubic capacity/cylinder	1752 cc/6 cylinder
hp/kW	46/34
Construction period	1931 – 1934
Number of produced cars	–

Alfa Romeo 6 C 2300 MM

From 1929 to 1939, Enzo Ferrari headed the racing department of Alfa Romeo. The victories that were regularly achieved by the manufacturing facility were also beneficial to vehicle development. In the middle of this era, the 6 C 2300 was launched, a model especially aimed at private drivers that was offered in the types Turismo, Gran Turismo and Pescara.

Model	Alfa Romeo 6 C 2300 MM
Cubic capacity/cylinder	2309 cc/6 cylinder
hp/kW	95/70
Construction period	1935 – 1939
Number of produced cars	–

Alfa Romeo 6 C

The 6 C was originally designed as a sports car for the Italian street race Mille Miglia. However, the manufacturer made the chassis available to famous coach makers because they were specialised in the production of labour-intensive special bodies. The 6 C shown here received such a body – produced by the Swiss coach maker Worblaufen.

Model	Alfa Romeo 6 C
Cubic capacity/cylinder	2309 cc/6 cylinder
hp/kW	95/70
Construction period	1935 – 1939
Number of produced cars	–

Alfa Romeo 6 C 2300 Mille Miglia

In the 1930s, it was common to construct the bodywork separate from the chassis including the propulsion technology, because self-supporting bodywork did not exist yet. The Alfa Romeo 6 C 2300 was also built in this way. The car shown here features a body especially created for racing.

Model	Alfa Romeo 6 C 2300 Mille Miglia
Cubic capacity/cylinder	2309 cc/6 cylinder
hp/kW	95/70
Construction period	1938 – 1939
Number of produced cars	–

Alfa Romeo 2300 Le Mans

In the 1930s, Alfa Romeo became the epitome of technically demanding and avant-garde cars. A reputation that was manifested by elegant designs such as the Alfa Romeo 2300 Le Mans. The experts agreed: "This Alfa Romeo is at least one decade ahead of its time in terms of style and aerodynamics".

Model	Alfa Romeo 2300 Le Mans
Cubic capacity/cylinder	2336 cc/8 cylinder
hp/kW	142/104
Construction period	1931 – 1934
Number of produced cars	–

Alfa Romeo 8 C 2300

The Alfa Romeo 8 C 2300 was equipped with one of the highest calibre engines ever developed for a sports car. Its eight cylinders were distributed among two identical four-cylinder blocks. This compressor car continued the many successes of its predecessors with victories including the Mille Miglia, the Targa Florio and Le Mans on several occasions.

Model	Alfa Romeo 8 C 2300
Cubic capacity/cylinder	2336 cc/8 cylinder
hp/kW	142/104
Construction period	1931 – 1934
Number of produced cars	

Alfa Romeo 8 C 2900 B

The 8 C 2900, the last Alfa Romeo built under the direction of engineer Vittorio Jano, differed from its predecessors in that it had a displacement of nearly tree litres. The force that powered the back wheels allowed the car to reach top speeds of up to 200 km/h, while the transmission ratio of the rear axle could be varied according to the purpose of the car.

Model	Alfa Romeo 8 C 2900 B
Cubic capacity/cylinder	2905 cc/8 cylinder
hp/kW	180/132
Construction period	1937 – 1939
Number of produced cars	30

Alfa Romeo 8 C 2900 B Speciale

In 1938, to really get the Alfa Romeo 8 C 2900 B Speciale going, the designers came up with something special: they equipped the 2.9-litre eight-cylinder engine with two so-called roots compressors. This increased its output and allowed the 8 C to reach a high speed of 240 km/h, which was quite amazing at the time.

Model	Alfa Romeo 8 C 2900 B Speciale
Cubic capacity/cylinder	2905 cc/8 cylinder
hp/kW	180/132
Construction period	1938 – 1939
Number of produced cars	–

Fiat 130 hp

Giovanni Agnelli, a former cavalry officer, was elected as the managing director of FIAT in 1902. He was considered to be a man of great ideas, and one year later the company already went public under his leadership. Soon the company also started to produce commercial vehicles and – very important for its image – race cars with large cubic capacity such as the 130 hp.

Model	Fiat 130 hp
Cubic capacity/cylinder	10.082 cc/4 cylinder
hp/kW	60/44
Construction period	1907
Number of produced cars	–

Fiat 508 S Balilla Sport

With a maximum speed of 110 km/h, the Fiat Balilla Sport was not a particularly fast vehicle, but it was certainly one of the most beautiful of its class. The coach maker Ghia designed the two-seater Spider body. Those who wanted to be exposed to even more fresh air could lower the small windshield. After the engine was reworked, the models built as of 1935 received an increase in power of 6 hp.

Model	Fiat 508 S Balilla Sport
Cubic capacity/cylinder	995 cc/4 cylinder
hp/kW	36/26
Construction period	1933 – 1936
Number of produced cars	113.145 (whole series)

Lancia Epsilon

In addition to progressive vehicles for everyday use, Vincenzo Lancia also produced cars with a sporty character, like the Epsilon. Lancia, who was already well-known in the racing world, always made a stir with progressive detail solutions. In 1919 for example, his cars already featured adjustable steering columns and four-wheel brakes.

Model	Lancia Epsilon
Cubic capacity/cylinder	4080 cc/4 cylinder
hp/kW	60/44
Construction period	1911 – 1913
Number of produced cars	–

Lancia Astura

"Design reveals the internal nature of a thing", Vincenzo Lancia once said. Faithful to this maxim he had the chassis of his cars covered in elegantly shaped metal sheets, while many famous coach makers worked for his company. With its luxury models of the 1920s and 1930s, the Dilambda and the Astura, Lancia established itself as the brand of choice of the rich, the beautiful and the famous.

Model	Lancia Astura
Cubic capacity/cylinder	2972 cc/8 cylinder
hp/kW	82/60
Construction period	1933 – 1939
Number of produced cars	–

SPA 30/40 HP

Established in 1906 by Giovanni Ceirano, "Società Ligure Piemontese Automobili" (in short SPA) aimed to primarily build highly competitive sports cars. At events such as the Targa Florio the strong SPA cars of the 7.7-litre class with their high cubic capacity gained many victories. For ambitious private drivers a smaller model was also available, the Type 30/40 hp.

Model	SPA 30/40 HP
Cubic capacity/cylinder	2658 cc/4 cylinder
hp/kW	40/29
Construction period	1912
Number of produced cars	

Spyker 60 HP Racer

The Dutch nationals Jakobus and Hendrik-Jan Spyker started building cars as early 1903. Their cars, which were always ahead of heir time, were successfully used in racing. In 1920/21 they set up a historical record by driving 30.000 kilometres non-stop in 36 days. However, due to the company's deteriorated financial situation, Spyker gave up automobile construction in 1925.

Model	Spyker 60 HP Racer
Cubic capacity/cylinder	8686 cc/6 cylinder
hp/kW	60/44
Construction period	1903
Number of produced cars	–

Auburn 12-160 V12

In 1919, US national Errett Lobban Cord took over the "Eckhart Carriage Company" with the goal of reorganising the company and converting it into a car factory. Under the trade name Auburn he wanted offer a luxury car that should cost only USD 1500. However, this price tag was obviously not high enough – an American luxury car was expected to cost tenfold this amount, which is why the brand was ignored.

Model	Auburn 12-160 V12
Cubic capacity/cylinder	6415 cc/12 cylinder
hp/kW	160/117
Construction period	1932 – 1936
Number of produced cars	–

Auburn Speedster

In 1936, Auburn had to give up automotive production due to financial mismanagement. In addition, the Cord and Duesenberg brands – both subsidiaries of the E. L. Cord Empire – also disappeared from the scene. Nevertheless, prior to the closure, Auburn was able to introduce one more compressor vehicle to the market, the Speedster, which is today considered to be one of the rarities of automotive history.

Model	Auburn Speedster
Cubic capacity/cylinder	4590 cc/8 cylinder
hp/kW	115 to 148/84 to 109
Construction period	1934 – 1936
Number of produced cars	–

Buick Y-Job

Automotive manufacturers frequently develop fascinating design-studies and concept-cars that are the stars and eye-catchers of motor shows. Unfortunately however, many of these highlights disappear into obscurity just as fast as they appear. In 1937, Buick deliberately presented a concept-car to the public for the first time, and the resonance was more than overwhelming.

Model	Buick Y-Job
Cubic capacity/cylinder	5200 cc/8 cylinder
hp/kW	141/103.2
Construction period	1937
Number of produced cars	unique piece

Cadillac Serie 90-V16

As a manufacturer of deluxe vehicles it was only natural for Cadillac to offer a big sixteen-cylinder along with its many eight-cylinder car models. While this Type 90 did not develop into a best-seller, since the target customers of this class of cars was very limited, it nevertheless contributed quite substantially to advancing the brand's image.

Model	Cadillac Serie 90-V16
Cubic capacity/cylinder	7063 cc/16 cylinder
hp/kW	185/135.5
Construction period	1930 – 1938
Number of produced cars	3250

Chrysler Imperial Speedster

Since his childhood, Walter P. Chrysler was preoccupied with automotive engineering. As an adolescent he already implemented his first construction plans. Chrysler was more than just a brilliant technician, he was also very well versed in the business side of this industry. At the age of 36 years he was promoted to a leading position at General Motors and given the task of bringing the Buick department into shape.

Model	Chrysler Imperial Speedster
Cubic capacity/cylinder	6308 cc/8 cylinder
hp/kW	135/99
Construction period	1932
Number of produced cars	–

Cord L 29

Errett Lobban Cord began his career in 1929 as a manufacturer of extremely luxurious cars. Insiders already knew his name, after all, among other achievements he was responsible for the development of the Duesenberg. The Cord's special feature was its transmission. The semi-automatic three-gear transmission of this front-wheel drive vehicle could be shifted by means of a tiny lever in the fascia.

Model	Cord L 29 series 1
Cubic capacity/cylinder	4893 cc/8 cylinder
hp/kW	125/92
Construction period	1929 – 1932
Number of produced cars	approx. 3600

Cord 812

Following the footsteps of the first Cord, the L 29, the second model of this American brand also joined the ranks of luxury cars. As opposed to the L 29 with its straight eight-cylinder engine, this model was fitted with a V8 engine, which could be equipped with a compressor to increase the performance output. One of the distinguishing features of the Cord 812 was the electromagnetic transmission, a type of semi-automatic transmission.

Model	Cord 812
Cubic capacity/cylinder	4730 cc/8 cylinder
hp/kW	175/128
Construction period	1937
Number of produced cars	2320 (all models)

Duesenberg J

Two descendants of German emigrants, Fred and August Duesenberg founded a company in the USA which specialised in building various luxury cars. Duesenberg models were always equipped with an eight-cylinder engine and their cubic capacity was on average about eight litres. When Duesenberg stopped manufacturing cars in 1937, they had by that time built a total of about 1300 cars.

Model	Duesenberg J
Cubic capacity/cylinder	6882 cc/8 cylinder
hp/kW	210/154
Construction period	1928 – 1937
Number of produced cars	approx. 480

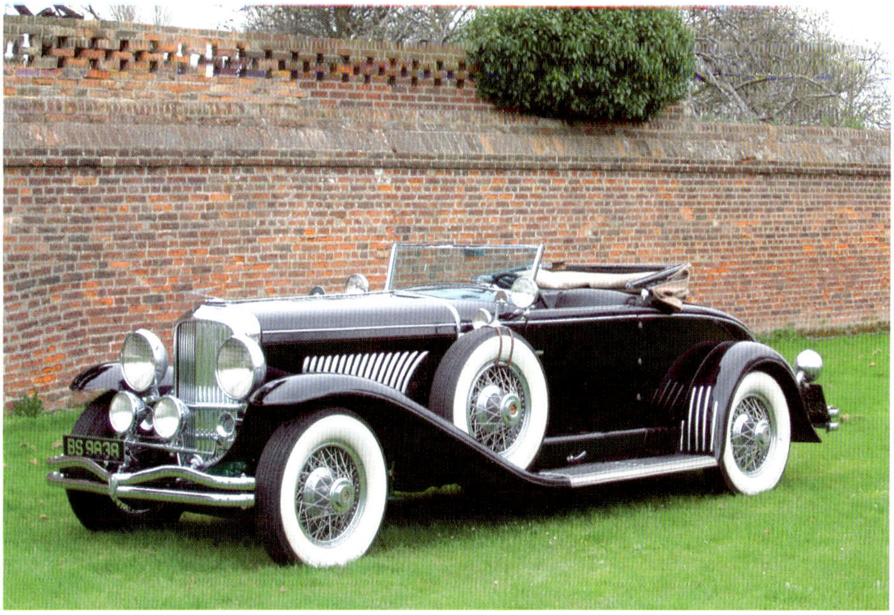

Duesenberg J

The advertising campaigns were just as interesting as the Duesenberg cars themselves. The campaigns refrained consciously, whenever possible, to use illustrations of the car, instead showing picturesque scenes of the surroundings which a Duesenberg-owner would enjoy, for example, a golf course. This was followed by the subtle message: "He drives a Duesenberg".

Model	Duesenberg J
Cubic capacity/cylinder	6882 cc/8 cylinder
hp/kW	210/154
Construction period	1928 – 1937
Number of produced cars	approx. 480

Duesenberg SJ

In addition to sedans, convertibles and roadsters, Duesenberg also created models that were primarily intended to demonstrate the performance capabilities of the brand in races. Utilising compression, the engine power was increased once more – enabling the eight-cylinder car, equipped with two overhead camshafts, to reach a speed of 208 km/h.

Model	Duesenberg SJ
Cubic capacity/cylinder	6882 cc/8 cylinder
hp/kW	320/235
Construction period	1933 – 1937
Number of produced cars	–

La Salle 303

General Motors Group created the La Salle brand as a cheaper alternative to the luxurious Cadillac. In the course of the development of the model, the La Salle models were fitted with increasingly powerful engines. When the straight eight-cylinder engine was replaced with a modern V8 engine, the La Salle acquired an undeniably sporty flair.

Model	La Salle 303
Cubic capacity/cylinder	4965 cc/8 cylinder
hp/kW	90/66
Construction period	1927 – 1928
Number of produced cars	–

Pierce Arrow

As the US company Pierce switched its scope of production more and more to automotive construction, the word "Arrow" replaced the old company name. This turned out to be a good idea, since the company wanted to particularly establish itself in the fast luxury car market. As of 1913, Pierce Arrow was the first automotive manufacturer to integrate the headlights into the wings.

Model	Pierce Arrow
Cubic capacity/cylinder	8577 cc/6 cylinder
hp/kW	75/55
Construction period	1919
Number of produced cars	–

Ruxton Roadster

In fact, the Ruxton Roadster was no in-house construction. Most of its prefabricated parts came from manufacturers who produced for the automotive industry for many years. Unfortunately, there was no demand for the Ruxton, and almost nobody wanted to purchase this luxury car. Shortly before the breakdown of the brand, the financiers managed to pull themselves out of the affair

Model	Ruxton Roadster
Cubic capacity/cylinder	5500 cc/8 cylinder
hp/kW	94/69
Construction period	1929 – 1931
Number of produced cars	–

Thomas Flyer 6-70

The US automotive brand Thomas existed from 1903 until 1918. The founder of the company, Erwin Ross Thomas, secured the biggest success of his career in 1906, when a Thomas Flyer 6-70 participating in the New York – Paris Long Distance Rally was the overall winner of this marathon drive through three continents covering more than 13,000 miles.

Model	Thomas Flyer 6-70
Cubic capacity/cylinder	12.800 cc/6 cylinder
hp/kW	72/53
Construction period	1906 – 1910
Number of produced cars	–

1945
1970

The passion for vintage cars – from affordable to exclusive

Cult objects of automotive history

In October of 1948, two new Jaguar sports cars, the XK 100 and the XK 120, were presented at the London Motor Show. However, it was still doubtful whether they would go into serial production. William Lyons, the head of Jaguar, wanted to first wait for the resonance of the audience before making the decision. This was no surprise since, even though the automotive press reported about luxury cars with increasing regularity, the mood was still rather negative in many European countries. Only some people with high incomes could afford expensive cars including sports cars – for the average consumer they remained an unattainable dream.

However, the enthusiasm which was caused by a Ferrari, a Jaguar or a Porsche 356 helped to improve this mood at least temporarily – this was because the sports cars were seen as messengers of the future and the bearers of hope at the same time. Not only was their appearance unique, but their interiors also made them unmistakable. The fascia alone with its many counters and levers was sometimes a regular work of sculptural art. Consequently, the advertisements for such vehicles had to be serious and at the same time innovative. Numerous photos and drawings demonstrated the utility of the cutting-edge technology or the many comfortable extras. Photographers and graphic artists were continuously creating a very positive image of these first sports cars of the post-war period.

In the 1950s, many automobile manufacturers started once again to

officially participate in international competitions with their serially produced sports cars. These events included for instance, the

Rallye Monte Carlo, the Critérium Neige et Glace, the Tour de Belgique or the Tour de Corse. The victories gained on snow and ice or on the asphalt racecourses were proof enough of the exceptional handling characteristics of a street sports car. But it was also possible to take the reverse course. For example, the legendary Mercedes-Benz gull-wing model was initially conceived as a pure racing car (300 SLR) and successfully entered in the 1952 Panamericana. The standard version of the Mercedes-Benz 300 SL was then conceived based on this winning car. The combination of the tubular space bar, which was taken over almost

identically from the racer, with the futuristic gull-wing doors inserted in the timelessly beautiful bodywork, remains unparalleled to this day. For the leaders of the era it was the status symbol of their success, for others it was an attainable desire – for most, however, it remained an inaccessible dream.

No other car came close to the unique status of the 300 SL. The competition did not even try to copy it. They rather considered it to be a driving force for innovation and came up with completely different models. BMW was committed to elegant eight-cylinder cars, Enzo Ferrari insisted on twelve cylinders, while on the other side of the Atlantic, Chevrolet launched the best selling Corvette. Paradoxically, in the land of giant cubic capacities and large hp engines, the Corvette was initially equipped with a mere six-cylinder engine.

With remarkable efficiency, the manufacturers serially provided their prestige objects with equipment that, until then, had only been available as options – if at all. This included power steering and power breaks as well as disc brakes and five-speed gearboxes. At that time already, all car manufacturers considered sports cars to be the benchmark that they had to live up to that spurred them on to achieve top performances simply to keep up with the competition.

Similar to the pre-war era, there was also a class of cars that looked stylish and fast but were powered by four-cylinder engines at most. They served the group of buyers which could "only" invest in a VW-Karmann-Ghia or an MG. Surprisingly, these "little ones" brought in a lot of money for the corporations as, from a business perspective, they were no more than mass products that could be produced at reasonable prices.

Auto Union 1000 Sp Cabrio

In 1957, Auto Union presented at the Frankfurt International Motor Show (IAA) the 1000 Sp, an elegant convertible whose most striking features were small rear fins. Even though the car did not have American dimensions, this style element suited it excellently. The advertisement portrayed models like the 1000 Sp as the sporty second car, which belonged to the lady of the house and was only driven by her.

Model	Auto Union 1000 Sp Cabrio
Cubic capacity/cylinder	980 cc/3 cylinder
hp/kW	55/40
Construction period	1958 – 1965
Number of produced cars	1640

Auto Union 1000 Sp Coupé

As an alternative to the convertible, Auto Union also produced a coupé version of the 1000 Sp. Both cars were technically identical and were equipped with a fixed-speed two-stroke engine. The surcharge list of the 1000 Sp was short, a lot of the options which other manufacturers offered for extra money (adjustable back supports, opening side windows, cigarette lighter, etc.) were already included in it as standard equipment.

Model	Auto Union 1000 Sp Coupé
Cubic capacity/cylinder	980 cc/3 cylinder
hp/kW	55/40
Construction period	1958 –1965
Number of produced cars	5000

BMW 507

"BMW has beaten the Italians in the class of high calibre sports-cars" rejoiced an auto magazine in 1955 reporting on the launch of the sports roadster BMW 507. The first concrete plans to construct the two-seater date back to 1954 already, and Albrecht Graf Goertz, the designer of this gem, simultaneously planned to present a bigger four-seat variation which would be built under the abbreviation 503.

Model	BMW 507
Cubic capacity/cylinder	3168 cc/8 cylinder
hp/kW	150/110
Construction period	1955 – 1959
Number of produced cars	254

BMW 507

After BMW had decided to construct the 507, it was certain that the shortened chassis of the big BMW sedan (Type 502–3.2 Super) could serve as the base for the roadster. The engine was also taken over from the sedan. However, in-spite of the slight increase in output, the engine was under-powered. Quite opposite to its expectations, BMW sold only 254 vehicles until it ceased production.

Model	BMW 507
Cubic capacity/cylinder	3168 cc/8 cylinder
hp/kW	150/110
Construction period	1955 – 1959
Number of produced cars	254

BMW 507 Vignale

The 507 also got its share of model development in spite of the low production numbers. This mostly consisted of technical improvements. Exterior changes were rather seldom, and with a few exceptions the roadster remained true to the initial model. One of these exceptions was the creation of a special body, the version shown here was created by Italian bodywork builder Vignale.

Model	BMW 507 Vignale
Cubic capacity/cylinder	3168 cc/8 cylinder
hp/kW	150/110
Construction period	1958
Number of produced cars	unique piece

BMW 503 Coupé

While the two-seater roadster BMW 507 was first presented in the USA, the bigger BMW 503 celebrated its debut at the Frankfurt International Motor Show (IAA) where it was presented to the audience in its two versions, a convertible and coupé. The press unanimously agreed "There is no pretentiousness, but a noble heritage behind the almost modest shapes, which are equivalent to the marvellous eight-cylinder engine …"

Model	BMW 503 Coupé
Cubic capacity/cylinder	3168 cc/8 cylinder
hp/kW	140/103
Construction period	1955 – 1960
Number of produced cars	412

BMW 503 Cabriolet

The distinguished elegance of the BMW 503 appealed exactly to the clientele that BMW had had in mind for this model – the high society. After all, one had to be rather wealthy to buy the coupé or the convertible, which cost DM 29.500 each. It is noteworthy that in the front one sat on comfortable single seats, while in the back there were two alternatives, either two dickey seats, or a diagonally (!) built-in single seat.

Model	BMW 503 Cabriolet
Cubic capacity/cylinder	3168 cc/8 cylinder
hp/kW	140/103
Construction period	1955 – 1960
Number of produced cars	412

BMW 3200 CS Bertone

After the debut of the models 503 and 507 there was a short break in the production of sporty vehicles by BMW. In 1961, the Bavarians presented a more potent, but very modestly styled model, the 3200 CS Bertone. Unfortunately, this car simply named "Bertone", also did not establish itself on the market, which is why BMW ceased its production in 1965.

Model	BMW 3200 CS Bertone
Cubic capacity/cylinder	3168 cc/8 cylinder
hp/kW	160/117
Construction period	1961 – 1965
Number of produced cars	587

BMW 2000 CS

In 1966, BMW introduced the elegant model 2000 CS, a coupé that was developed in house, but built outside by Karmann in Osnabrück. Specifically, the designers dedicated a lot of care to the exclusive interior. While the two-litre models were equipped with a four-cylinder engine, the cars with 2.8 litres of cubic capacity and upwards received a six-cylinder engine.

Model	BMW 2000 CS
Cubic capacity/cylinder	1990 cc/4 cylinder
hp/kW	120/88
Construction period	1966 – 1969
Number of produced cars	–

BMW 3.0 CSi

When the BMW 2000 CS Coupé (four cylinders) had matured in 1968 into the 2800 CS (six cylinders), it had received a substantially more balanced form, thanks to the extended hood and other slight optical modifications. The 3.0 CSi, whose production started in 1971, was even equipped with a draught fuel injection engine. The 220 hp transmitted to the rear axle provided a maximum speed of 220 km/h.

Model	BMW 3.0 CSi
Cubic capacity/cylinder	2985 cc/6 cylinder
hp/kW	220/161
Construction period	1971 – 1975
Number of produced cars	–

BMW 1600 GT

In addition to small cars (Goggomobil), Hans Glas also built midrange and high-end cars. The Glas 1700 GT remained in production up to the breakdown of the Glas company. The production of the model continued in 1966 under the direction of the new owner (BMW). BMW modified the front part of the coupé, integrated a smaller engine and brought out the model as a BMW 1600 GT on the market.

Model	BMW 1600 GT
Cubic capacity/cylinder	1573 cc/4 cylinder
hp/kW	105/77
Construction period	1966 – 1968
Number of produced cars	1255

DKW Monza

In 1956 the DKW Monza set the world record for 4000 miles, 5000 miles and 10.000 kilometres in the class of up to 1100 cubic centimetres. Many of the coupés made in small series with plastic bodywork went to the USA. The fact that only 230 units were built, was due to the fact that the Auto Union GmbH already had planned a hot new project to introduce to the sports car market – The Auto Union 1000 Sp which was released in 1957.

Model	DKW Monza
Cubic capacity/cylinder	980 cc/3 cylinder
hp/kW	44/33
Construction period	1956 – 1958
Number of produced cars	230

Ford Capri 2000 GT

With the Capri, Ford had undoubtedly created a trend-setting car, which not only made the hearts of young drivers beat faster. The sporty model was introduced to the market with an interesting price-performance ratio, and quickly worked its way up, thanks to a well-assorted model range. In 1974, the Capri had to make room for the second generation.

Model	Ford Capri 2000 GT
Cubic capacity/cylinder	1988 cc/6 cylinder
hp/kW	90/66
Construction period	1969–1972
Number of produced cars	approx. 784.000

Glas 2600 V8/BMW 3000 V8

The Glas 2600 V8 introduced in 1965 was a luxury car with a body sketched by the designer Pietro Frua. In 1966, Hans Glas – the small car manufacturer who became famous after producing the Goggomobil – had to sell his ailing enterprise to BMW. Under the direction of BMW the car was called BMW 3000 V8 and remained in production until 1968.

Model	BMW 3000 V8
Cubic capacity/cylinder	2982 cc/8 cylinder
hp/kW	160/117
Construction period	1966 – 1968
Number of produced cars	698

Mercedes-Benz 300 SL

The year 1951 was not only the year Daimler-Benz gained two world titles in the Formula 1, but it also was the birth year of an immortal cult car – the mythical SL. No other abbreviation has attained similar charismatic shine like the model name SL ("Sporty" and "Light").

Model	Mercedes-Benz 300 SL
Cubic capacity/cylinder	2996 cc/6 cylinder
hp/kW	215/158
Construction period	1954 – 1957
Number of produced cars	1400

Mercedes-Benz 300 SL

Although the first SL – a racing car model – was only based on the salon W 186 (the famous "three hundred"), the sports car derived from it (Type SLR) was immediately successful. Rudolf Uhlenhaut who had provided the stimulus for this car said: "We took the standard engine of the 300 and built a tubular frame with an aluminium body around it."

Model	Mercedes-Benz 300 SL
Cubic capacity/cylinder	2996 cc/6 cylinder
hp/kW	215/158
Construction period	1954 – 1957
Number of produced cars	1400

Mercedes-Benz 300 SL

In 1952, after the double victory of Karl Kling and Hans Klenk as well as Hermann Lang and Erwin Grupp and on the occasion of the Panamericana, the gull-wing door coupé finally moved to the centre of media interest. In the same year, another 300 SL participated in the Mille-Miglia and finished in second place, while at the 24 hours of Le Mans and on the Nürburgring the 300 SL cars were among the quickest.

Model	Mercedes-Benz 300 SL
Cubic capacity/cylinder	2996 cc/6 cylinder
hp/kW	215/158
Construction period	1954 – 1957
Number of produced cars	1400

Mercedes-Benz 300 SL

The idea to also build the 300 SL as a street version came from Max Hoffman. Hoffmann, who dealt in European vehicles in the USA, assumed that he would be able to sell 1000 units of the street legal model. His arguments were convincing, and thus the street version of 300 SL saw the light of the automotive world in February 1954 for the first time at the New York Motor Show.

Model	Mercedes-Benz 300 SL
Cubic capacity/cylinder	2996 cc/6 cylinder
hp/kW	215/158
Construction period	1954 – 1957
Number of produced cars	1400

Mercedes-Benz 300 SL

A six-cylinder high performance engine, which produced 210 hp, was developed especially for the 300 SL. Equipped with a sports camshaft and slightly higher compression, the straight engine was capable of an output in excess of 215 hp, which corresponded to the, at that time, inconceivably high value of 71.5 hp per each litre of cubic capacity. The engine revolutions also caused a great deal of admiration – 6600 rpm were labelled as maximum.

Model	Mercedes-Benz 300 SL
Cubic capacity/cylinder	2996 cc/6 cylinder
hp/kW	215/158
Construction period	1954 – 1957
Number of produced cars	1400

Mercedes-Benz 300 SL

The principle of the tubular space frame, on which 300 SL is based, comes from the field of aeroplane construction, because constructions of this kind are known to be especially sturdy. Additionally, the use of the tubular space frame permitted another impressive feature, the gull-wing doors. However, the frame construction of the roadster had to be slightly modified, to provide for a comfortable and easy entry.

Model	Mercedes-Benz 300 SL
Cubic capacity/cylinder	2996 cc/6 cylinder
hp/kW	215/158
Construction period	1954 – 1957
Number of produced cars	1400

Mercedes-Benz 300 SL Roadster

In 1957, the gull-wing door was removed from 300 SL roadsters. Starting in 1958, Daimler-Benz also offered the car with an optional hard top. SL owners should, naturally, be able to enjoy the driving experience in any sort of weather. Celebrities like Zsa Zsa Gabor, the Duke of Edinburgh, Shah Reza Pahlevi or Elvis Presley, alongside private individuals, were among the enthusiasts who drove a SL.

Model	Mercedes-Benz 300 SL Roadster
Cubic capacity/cylinder	2996 cc/6 cylinder
hp/kW	215/158
Construction period	1957 – 1963
Number of produced cars	1858

Mercedes-Benz 300 SL Roadster

Compared to the gull-wing door model, the roadster profited in the course of model development from various improvements – a new single-joint swing axle with low pivot point and a buffer spring replacing the original dual-joint axle, which made the car easier to handle at top limits of performance, regardless of the driver's skill. As of 1961, the roadster featured disc brakes on all four wheels.

Model	Mercedes-Benz 300 SL Roadster
Cubic capacity/cylinder	2996 cc/6 cylinder
hp/kW	215/158
Construction period	1957–1963
Number of produced cars	1858

Mercedes-Benz 300 SL Roadster

A leading American automobile magazine once wrote in describing the 300 SL "... the sports car of the future has become a reality!" Unfortunately, this reality ended in February 1963 when Daimler-Benz discontinued the production of the 300 SL. Up to that time, the sports car had been built a total of 3258 times, of which 1858 times were roadster models. Its successor, the "Pagoden SL", had to live up to this impressive legacy.

Model	Mercedes-Benz 300 SL Roadster
Cubic capacity/cylinder	2996 cc/6 cylinder
hp/kW	215/158
Construction period	1957 – 1963
Number of produced cars	1858

Mercedes-Benz 190 SL

The US market had been an interesting market for sports car manufacturers for a long time. Max Hoffmann, a German American, who dealt with German cars in the USA since 1946, knew exactly what was demanded there. He advised Daimler-Benz to definitely launch the 190 SL model. The 190 SL subsequently sold excellently in the USA, proving how intuitive he was.

Model	Mercedes-Benz 190 SL
Cubic capacity/cylinder	1897 cc/4 cylinder
hp/kW	107/77
Construction period	1955 – 1963
Number of produced cars	25.881

Mercedes-Benz 190 SL

In 1954, the automotive press reported regularly about the 190 SL, however, it was still unclear when the car would be available in the market. Insiders knew the reason – there were some weak points that constantly delayed the start of the series. Additionally, they knew that the engineers were only given five months left for the development of the 190 SL after the management board discussions …

Model	Mercedes-Benz 190 SL
Cubic capacity/cylinder	1897 cc/4 cylinder
hp/kW	107/77
Construction period	1955 – 1963
Number of produced cars	25.881

Mercedes-Benz 190 SL

While searching for a suitable engine for the new 190 SL it became clear that the most logical solution, by far, would be construct a new one. Unfortunately, the four cylinder engine turned out to be unruly in driving. The problem was solved only by complex experimentation with different carburettors.

Model	Mercedes-Benz 190 SL
Cubic capacity/cylinder	1897 cc/4 cylinder
hp/kW	107/77
Construction period	1955 – 1963
Number of produced cars	25.881

Mercedes-Benz 190 SL

The moment prospective buyers and automobile enthusiasts had been long waiting for arrived: On the occasion of the Geneva Motor Show in 1955, Daimler-Benz finally presented the production-ready model of the 190 SL. When one considered the prospects it was obvious that the 190 SL had not been purely conceived as a sports car like the 300 SL. Rather, the 190 SL two-seater belonged to the class of sportily elegant touring and everyday-use vehicles.

Model	Mercedes-Benz 190 SL
Cubic capacity/cylinder	1897 cc/4 cylinder
hp/kW	107/77
Construction period	1955 – 1963
Number of produced cars	25.881

Mercedes-Benz 230 SL

In 1963 Daimler-Benz presented the 230 SL at the Geneva Motor Show – the official successor of the 300 SL. Its appearance was rather unusual, because the roof – or the hard top to be more precise – was lowered pagoda-shaped over the centre of the car. However, it became widely known that the second SL generation was an excellent touring car, which could easily reach a speed of 200 km/h.

Model	Mercedes-Benz 230 SL
Cubic capacity/cylinder	2306 cc/6 cylinder
hp/kW	150/110
Construction period	1963 – 1971
Number of produced cars	48.912

Mercedes-Benz 250 SL

The equally strong 250 SL replaced the 230 SL in the beginning of 1967. Eleven months later, after producing only 5196 units, the 280 SL (170 hp) was already available for test-drives at dealerships. As a matter of fact, nearly 70 percent of all cars were exported. Most went to the USA and were predominantly equipped with power steering and automatic transmissions.

Model	Mercedes-Benz 250 SL
Cubic capacity/cylinder	2496 cc/6 cylinder
hp/kW	150/110
Construction period	1967–1968
Number of produced cars	5196

Mercedes-Benz 280 SL

According to factory specifications, the Types 230 SL and 280 SL were the first sport cars in the world which featured a "safety body with a rigid passenger cabin and crush zones in front and rear." Additionally, the roof construction, in terms of the ease of use, was regarded in the 1960s by at least one professional magazine as the "fastest roof in the world".

Model	Mercedes-Benz 280 SL
Cubic capacity/cylinder	2778 cc/6 cylinder
hp/kW	170/125
Construction period	1968 – 1971
Number of produced cars	–

Opel Rallye Kadett

From its initial concept, the Rallye Kadett was based on the coupé variation of the Kadett B series launched a year before. In the beginning it was modestly equipped with an improved performance 1.1-litre engine. However in 1968, and because of the demand for more performance, Opel introduced a model with 1.9 litres, a modification that predestined the Rallye Kadett for racing.

Model	Opel Rallye Kadett
Cubic capacity/cylinder	1897 cc/4 cylinder
hp/kW	90/66
Construction period	1967 – 1973
Number of produced cars	–

Opel GT

In 1965, Opel showed a coupé derived by the Kadett, which was equipped with a 1.9-litre engine at the Frankfurt International Motor Show. Over the course of the next years this concept car was further developed, until it celebrated its première, in the end, in 1968 under the name Opel GT. The bodywork of the GT was made, incidentally, in France, while the assembly took place at the factory in Bochum.

Model	Opel GT
Cubic capacity/cylinder	1897 cc/4 cylinder
hp/kW	90/66
Construction period	1968 – 1973
Number of produced cars	103.373

Opel GT/J

For two years, from 1968 to 1970, a 60 hp variation of the 90 hp GT was also produced as an alternative model. The top speed of 155 km/h was unfortunately not enough for a GT, which is why it sold very badly. Opel had more success with the GT/J. While it correlated technically to the 90 hp model, the chrome plated mouldings and the luxurious interior had to be dropped for cost reasons.

Model	Opel GT/J
Cubic capacity/cylinder	1897 cc/4 cylinder
hp/kW	90/66
Construction period	1968 – 1973
Number of produced cars	103.373

Porsche 356

Ferdinand "Ferry" Porsche (who passed away at the age of 88, on the 27th of March 1998) and his employees had begun the project with the development number 356 at Gmünd in Carinthia, Austria, where the company moved during the war. The first design drawings were made on 17th July 1947, and the provincial government of Carinthia, at that time, gave a single approval for the licensing of the car on 8th June 1948.

Model	Porsche 356
Cubic capacity/cylinder	1131 cc/4 cylinder
hp/kW	40/29
Construction period	1948
Number of produced cars	unique piece

Porsche 356 Speedster

As Dr. Ferdinand Porsche ("Ferrys" father) saw the first Type 356 with the chassis number 356 001, he is quoted to have said: "I would not have built even one screw differently." Following the number 1, a small series of 52 cars was built in Gmünd. In Stuttgart, a total of 78,000 units of the constantly further developed Type 356 were built from 1950 to 1965.

Model	Porsche 356 Speedster
Cubic capacity/cylinder	1582 cc/4 cylinder
hp/kW	75/55
Construction period	1955 – 1959
Number of produced cars	–

Porsche 356 A Cabriolet

The first sports car of the Porsche brand was powered by a modified 1.1-litre Volkswagen engine. This air cooled engine which otherwise was only used in the VW Beetle, provided the Porsche 356 with a top speed of 135 km/h. In the beginning the power output was a modest 35 hp. The engine power was continuously increased as part of the model development.

Model	Porsche 356 A Cabriolet
Cubic capacity/cylinder	1588 cc/4 cylinder
hp/kW	115/85
Construction period	1959 – 1960
Number of produced cars	–

Porsche 356 B Cabriolet

Porsche's success in racing was also initiated with the Type 356. In July 1948, the car with the chassis number 356 001 was only a few weeks old, when it achieved victory in its class in a city race in Innsbruck. The success in racing prompted Porsche to generously expand the model range (including the cabriolet, the speedster and the coupé), and to add to it increasingly powerful models.

Model	Porsche 356 B Cabriolet
Cubic capacity/cylinder	1582 cc/4 cylinder
hp/kW	60/44
Construction period	1959 – 1962
Number of produced cars	–

Porsche 356 C Cabriolet

With the introduction of the model range 356 C in August 1963, there was plenty for the trade journalists to write about – the 356 C profited from numerous technical innovations. Among other modification, this easy to handle sports car was now equipped with disc brakes. To improve the driving comfort, the rear axle suspension was softened a little bit, and it was also complemented with a stabiliser bar.

Model	Porsche 356 C Cabriolet
Cubic capacity/cylinder	1582 cc/4 cylinder
hp/kW	75/55
Construction period	1963 – 1965
Number of produced cars	–

Porsche 356 C Coupé

After only one month of the Porsche 356 debut, a rumour circulated that the days of this model were numbered. Insiders knew why – a new sports car, the Porsche 911, was just about to be launched. However, to the satisfaction of all Porsche fans, the 356 was not yet dropped from the model range. The company concentrated on the production of the variations with 75 hp and 90 hp that continued to roll off the conveyor belts.

Model	Porsche 356 C Coupé
Cubic capacity/cylinder	1582 cc/4 cylinder
hp/kW	75/55
Construction period	1963 – 1965
Number of produced cars	–

Porsche 356 C Carrera Coupé

Until the end of production of the 356 in 1965, about 78.000 buyers from all over the world had found pleasure in Ferry Porsche's car. Other sports car models, headed by the 911, advanced the brand to one of the most famous automobile manufacturers, known for successful designs as well as groundbreaking and reliable technology, two characteristics that continue to this day.

Model	Porsche 356 C Carrera Coupé
Cubic capacity/cylinder	1966 cc/4 cylinder
hp/kW	130/96
Construction period	1963 – 1965
Number of produced cars	–

Porsche 901

Porsche presented the new 901, which was later renamed 911, at the Frankfurt International Motor Show (IAA) in 1963. The successor of the extremely successful Type 356 was designed by Ferdinand A. Porsche, the oldest son of the head of the company Ferry Porsche. As opposed to the 356 the car featured more interior space and for the first time a six-cylinder engine instead of the customary four-cylinder engine.

Model	Porsche 901
Cubic capacity/cylinder	1991 cc/6 cylinder
hp/kW	130/96
Construction period	1963
Number of produced cars	prototype

Porsche 911 2.0

When the Porsche 911 went into production in 1964, its retail price was around DM 21,900. For this amount, the customer received a pure bred sports car with an output of 130 hp and a maximum speed of 210 km/h. As with its predecessor, the Porsche 356, an air-cooled boxer engine was again placed in the rear. This time, however, instead of using a four-cylinder, a six-cylinder engine was installed.

Model	Porsche 911 2.0
Cubic capacity/cylinder	1991 cc/6 cylinder
hp/kW	130/96
Construction period	1964 – 1968
Number of produced cars	–

Porsche 912

The demand for a more affordable Porsche led to the presentation of the 912 model in 1965. Its 90-hp four-cylinder engine was still derived from the 356 model range. In conjunction with basic interior furnishing, the retail price of the 912 was DM 16,250. However, the 912 was, of course, not as vicious as the 911, as it required 13.5 seconds to accelerate from 0 to 100 km/h and only reached a speed of up to 183 km/h.

Model	Porsche 912
Cubic capacity/cylinder	1582 cc/4 cylinder
hp/kW	90/66
Construction period	1965 – 1968
Number of produced cars	–

Porsche 911 Targa

In September 1965, Porsche introduced the 911 Targa, a new body variation including a permanent roll bar, detachable roof and initially a removable plastic rear window. Those who wanted to buy this fresh air vehicle had to be very patient, because the production began only substantially later, in December 1966. In terms of model development, the car, which was offered in several output variations, was modelled after the coupé.

Model	Porsche 911 Targa
Cubic capacity/cylinder	1991 cc/6 cylinder
hp/kW	110/81
Construction period	1966 – 1969
Number of produced cars	–

Porsche Type 718/8 RS Spyder

In 1962, the 718/8 could be seen on many racetracks of Europe and Porsche even showed the car in distant California. The première of the car already took place in 1961 while it was still equipped with a four-cylinder engine. Many insights that were gained from this model with a top speed of 260 km/h were later applied to the development of the 904 GTS Coupé.

Model	Porsche Type 718/8 RS Spyder
Cubic capacity/cylinder	1981 cc/8 cylinder
hp/kW	210/154
Construction period	1962
Number of produced cars	–

Porsche 904 GTS

The 904 GTS , also known as the Carrera GTS, was Porsche's first vehicle that was equipped with a synthetic body. To be able to start competing in the GT class, 100 copies of the model had to be built in the street version (with a four-cylinder engine). As usual for Porsche, the majority of the cars (43 units) with a top speed of 263 km/h was exported to the USA.

Model	Porsche 904 GTS
Cubic capacity/cylinder	1966 cc/4 cylinder
hp/kW	180/132
Construction period	1963 – 1965
Number of produced cars	100

Veritas 90 SPC

The small sports car manufacturer Veritas was originally headquartered in Messkirch in the German region of Baden, but was moved to the Nürburgring in 1951. Here they produced in minute volumes different sports car models, which were mainly equipped with BMW technology. As expected by the exclusive nature of a Veritas, the sports cars from the Eiffel were not exactly among the most affordable products around.

Model	Veritas 90 SPC
Cubic capacity/cylinder	1988 cc/6 cylinder
hp/kW	100/73
Construction period	1949 – 1950
Number of produced cars	–

Volkswagen Karmann-Ghia

The idea to build a sporty-looking two-seater did not come to the Karmann factory by chance – after all, they already produced the VW Beetle convertible. In their opinion, an increasing number of citizens slowly became prosperous and some even considered the acquisition of a second car – so why should this not be a model with a sporty "packaging"?

Model	Volkswagen Karmann-Ghia
Cubic capacity/cylinder	1192 cc/4 cylinder
hp/kW	30/22
Construction period	1955 – 1960
Number of produced cars	362.585 (coupés)

Volkswagen Karmann-Ghia

Wilhelm Karmann, boss of the Karmann factory in Osnabrück, had achieved his goal in 1953 – the Volkswagen Group gave him the "green light" for the development of a little sports car based on the beetle. When the prototype model of the car was finally shown to the press, everyone agreed that the pretty coupé with German technology and Italian design had to definitively go into production.

Model	Volkswagen Karmann-Ghia
Cubic capacity/cylinder	1192 cc/4 cylinder
hp/kW	34/25
Construction period	1959 – 1960
Number of produced cars	80.881 (convertibles)

Volkswagen Karmann-Ghia

While the Karmann-Ghia Coupé, produced since 1955, enjoyed increasing popularity, the conveyor belts for the convertible version only began to operate two years later. In the meantime, the factory considered also optionally equipping the car with a hard top. Unfortunately, this possibility was rejected again. However, the prototype produced for this purpose still exists today.

Model	Volkswagen Karmann-Ghia
Cubic capacity/cylinder	1192 cc/4 cylinder
hp/kW	30/22
Construction period	1957
Number of produced cars	unique piece

Volkswagen Karmann-Ghia

The good thing about the Karmann-Ghia was certainly its close relationship to the VW Beetle with which it shared many common characteristics. Every time the Beetle was modified as part of model development, the sporty-looking offshoot benefited as well. The career of the best-selling Karmann-Ghia ended only after nearly 20 years of construction, in July 1974. The last car that came off the conveyor belt was a white convertible.

Model	Volkswagen Karmann-Ghia
Cubic capacity/cylinder	1584 cc/4 cylinder
hp/kW	50/37
Construction period	1971–1974
Number of produced cars	

VW-Porsche 914

In addition to decades of development activities for the Volkswagen plant, a closer collaboration with VW took place once again when Ferry Porsche decided to offer a sports car below the price range of the 911. Because the engineers in Wolfsburg as well as in Stuttgart-Zuffenhausen were persuaded of the mid-engine car proposal, the project was realised under the model name 914.

Model	VW-Porsche 914
Cubic capacity/cylinder	1679 cc/4 cylinder
hp/kW	80/59
Construction period	1969 – 1975
Number of produced cars	115.646

VW-Porsche 914/6

In 1969, the Type 914 was presented as a VW Porsche to the general public. In terms of construction, it consisted of a mid-engine sports car with a removable roof that was initially equipped with a VW injection engine (four cylinders and 1.7 litres of cubic capacity). A more vicious alternative was the 914/6 under whose bonnet a modified two litre Porsche engine with six cylinders could be found.

Model	VW-Porsche 914/6
Cubic capacity/cylinder	1991 cc/6 cylinder
hp/kW	110/81
Construction period	1969 – 1972
Number of produced cars	3332

AC Ace Bristol

In the early 1950s, John Tojeiro, a race car designer, developed this elegant sports car that was built by the British company AC as of 1953. At first, the 86-hp model AC Ace debuted followed in 1956 by the AC Ace Bristol that was equipped with a Bristol engine. Thanks to its lightweight construction (aluminium bodywork) the Bristol variation reached 200 km/h.

Model	AC Ace Bristol
Cubic capacity/cylinder	1971 cc/6 cylinder
hp/kW	130/96
Construction period	1956 – 1963
Number of produced cars	approx. 465

AC Aceca

AC offered an elegant coupé model, the Aceca, for car enthusiasts who were looking for a sports car with a closed roof. The Aceca, which was essentially the closed version of the Ace Bristol convertible, however, also profited due to its design from a practical boot lid. As of 1957 incidentally, the Aceca was also available with front disc brakes at a surcharge.

Model	AC Aceca
Cubic capacity/cylinder	1971 cc/6 cylinder
hp/kW	130/96
Construction period	1954 – 1963
Number of produced cars	approx. 170

AC Ace

After construction small tricycle cars for many years, the British company AC (Auto Carrier) surprisingly introduced its first sports car in 1953. Based on the positive feedback to this "AC Ace", it soon became the production base for a new vehicle generation. Until 1961, AC equipped the sporty vehicle with a Bristol six-cylinder engine, while later versions received a V8 engine by Ford.

Model	AC Ace
Cubic capacity/cylinder	1991 cc/6 cylinder
hp/kW	127/93
Construction period	1960 – 1961
Number of produced cars	–

AC Shelby Cobra

In 1962, the native Texan Carroll Shelby (at that time a racing driver) combined a chassis built by the British company AC with an American Ford V8 engine, and called the result of this synthesis Cobra. This car, which was subsequently built in a small batch, performed excellently not only on the street, but also on the racetrack – the highly tuned variations of this sports car regularly scored victories.

Model	AC Shelby Cobra
Cubic capacity/cylinder	7033 cc/8 cylinder
hp/kW	359/257
Construction period	1963 – 1969
Number of produced cars	–

AC Cobra 427

For sports car experts, the Cobra 427 remains the most brutal car ever released on the road. To be able to keep this power-loaded racer on the ground, its chassis was regularly revised and optimised. The manufacturer could guarantee a maximum speed of 240 km/h for the smallest engine option already – with no upper limit.

Model	AC Cobra 427
Cubic capacity/cylinder	6997 cc/8 cylinder
hp/kW	425/311
Construction period	1965 – 1968
Number of produced cars	410

Allard J2X Le Mans

The Briton Sydney Allard – who came in third in the 1950 Le Mans race – launched some fascinating sports cars from 1946 to 1955. As he did with most of his models, he also put on the J2X a body that was reduced to the basics to keep the car very light. Powerful engines by Cadillac, Chevrolet or Chrysler allowed the Allard cars to reach speeds of up to 200 km/h.

Model	Allard J2X Le Mans
Cubic capacity/cylinder	5240 cc/8 cylinder
hp/kW	162/119
Construction period	1952 – 1954
Number of produced cars	–

Aston Martin DB 2

In 1947, David Brown, a British industrialist, took over the Aston Martin plant. He restructured the financially ailing enterprise and launched a new series of cars only one year later. In 1948, he introduced the DB 1 (DB for David Brown), a rather baroque-looking sports car. Its successor, the DB 2 with a more modern silhouette was substantially more elegant.

Model	Aston Martin DB 2
Cubic capacity/cylinder	2580 cc/6 cylinder
hp/kW	108/79
Construction period	1951 – 1953
Number of produced cars	–

Aston Martin DB 2

In addition to the Aston Martin, David Brown also reorganised the noble brand Lagonda. This provided him with the rights to an engine construction that Lagonda had originally developed for Bentley. David Brown used this six-cylinder engine with two overhead camshafts from that time on for his DB 2 model. The body of the DB 2 originated, incidentally from the firm of Tickford – which was owned by David Brown.

Model	Aston Martin DB 2
Cubic capacity/cylinder	2580 cc/6 cylinder
hp/kW	108/79
Construction period	1951 – 1953
Number of produced cars	–

Aston Martin DB 2/4

It was only a small, but effective feat – in 1953, Aston Martin equipped a DB 2 chassis with a bigger body which provided more space in the rear. The result was definitely worthwhile – the originally two-seater was changed into a four-seater. As part of the model development, the rear window of the DB 2/4 was shortly thereafter integrated into the boot lid.

Model	Aston Martin DB 2/4
Cubic capacity/cylinder	2580 cc/6 cylinder
hp/kW	127/93
Construction period	1953 – 1954
Number of produced cars	–

Aston Martin DB 2/4 Mk III

In 1957, Aston Martin presented the definitive perfection of the DB 2/4 – the Mk III. Compared to its predecessors, the engine of the Mk III received a newly-constructed cylinder-head. In addition, the car featured a slightly modified chassis and disc brakes by request. In the course of model development, the interior was further upgraded and the fascia was redecorated with elements including screened instruments.

Model	Aston Martin DB 2/4 Mk III
Cubic capacity/cylinder	2922 cc/6 cylinder
hp/kW	164/120
Construction period	1956 – 1957
Number of produced cars	–

Aston Martin DB 2/4

In 1957, Aston Martin introduced further developments to the DB 2/4 model. In the DB 2/4 Mk III version, the nicely styled sports car benefited from the strength of a modern three-litre engine. As a rule, the car that could optionally be equipped with disc brakes was produced only as a coupé or convertible. A spider version of only three copies was built by Touring in Italy.

Model	Aston Martin DB 2/4
Cubic capacity/cylinder	2922 cc/6 cylinder
hp/kW	164/120
Construction period	1956 – 1957
Number of produced cars	3

Aston Martin DB 4 GT

With the DB 4 model Aston Martin implemented the idea of a sporty luxury touring car – and successfully so! The DB 4 remained part of the product range until 1963 (all model types). During this time it underwent five model development measures. The GT version that had a maximum speed of 250 km/h with a slightly shorter wheel base was conceived as a two-seater and had a higher output than the standard model.

Model	Aston Martin DB 4 GT
Cubic capacity/cylinder	3670 cc/6 cylinder
hp/kW	302/222
Construction period	1959 – 1960
Number of produced cars	75

Aston Martin DB 4 GT Zagato

Even in those days, the DB 4 GT was among the cars that were only rarely seen. Aston Martin introduced an even rarer version of the car, the DB 4 GT Zagato. This model was equipped with a body of Italian coach maker Zagato and was manufactured only 25 times. Today, the Zagato is among the most sought-after car models. In comparison: a total of 1119 units were produced of the standard DB 4 model.

Model	Aston Martin DB 4 GT Zagato
Cubic capacity/cylinder	3670 cc/6 cylinder
hp/kW	314/231
Construction period	1960 – 1961
Number of produced cars	25

Aston Martin DB 5

When David Brown took over Aston Martin in 1947, the company's first step was to develop the famous DB series. After the DB 1, DB 2 and DB 4 the noble brand launched another object of the desire – the DB 5. This version became primarily known through the James Bond movies. Not only "007" drove an Aston Martin – the DB 5 was of course also available for "normal consumers".

Model	Aston Martin DB 5
Cubic capacity/cylinder	3995 cc/6 cylinder
hp/kW	282/207
Construction period	1963 – 1965
Number of produced cars	1063

Aston Martin DB 6 Mk II

The successor of the DB 5, the DB 6, primarily differed from its predecessor by a somewhat longer wheel base and a slightly revised body. Aston Martin also increased the total length of the new model, which resulted in an even more harmonious shape. In addition, a so-called trailing edge in the rear highlighted the sporty character of the DB 6.

Model	Aston Martin DB 6 Mk II
Cubic capacity/cylinder	3995 cc/6 cylinder
hp/kW	286/210
Construction period	1965 – 1970
Number of produced cars	1755

Aston Martin DB 6 Volante

To further perfect the DB 6, and as part of model development, the company equipped the car with power steering. In addition, the list of extras included air-conditioning, while fans of convertible cars were also offered the DB 6 as a Volante, which is what open models were called at Aston Martin. The DB, with a maximum speed of 250 km/h incidentally also turned into a best-seller in the US market.

Model	Aston Martin DB 6 Volante
Cubic capacity/cylinder	3995 cc/6 cylinder
hp/kW	286/210
Construction period	1965 – 1970
Number of produced cars	1755

Aston Martin DBS V8

In 1967, the model DBS, a rather conservatively styled car, had its debut. The coupés built until 1969 were equipped with six-cylinder engines, while later versions received a V8 engine with 2x2 overhead camshafts. As opposed to the DB 6, the company propagated the DBS as a full four-seater – with a total length of 459 centimetres.

Model	Aston Martin DBS V8
Cubic capacity/cylinder	5340 cc/8 cylinder
hp/kW	340/249
Construction period	1969 – 1972
Number of produced cars	405

Austin-Healey 100

In 1951, the Briton Donald Healey showed the prototype of an automobile at the London Motor Show. The design was also received with enthusiasm by the board of directors of the Austin car manufacturers, because Austin had been planning the release of a sports car for a long time. Healey recognised the chance of having his endeavour implemented by Austin and offered them the production rights.

Model	Austin-Healey 100
Cubic capacity/cylinder	2660 cc/4 cylinder
hp/kW	91/67
Construction period	1952 – 1956
Number of produced cars	approx. 12.900

Austin-Healey 100/6

By experimenting with a powerful six-cylinder, Donald Healey had the idea of offering an improved version of his sports car, which until then had only been equipped with a four-cylinder engine. In 1956, the car with the added driving pleasure was launched under the name 100/6. In addition to improved driving conditions, the newcomer also surprisingly offered jump seats in the back.

Model	Austin-Healey 100/6
Cubic capacity/cylinder	2639 cc/6 cylinder
hp/kW	103/76
Construction period	1956 – 1959
Number of produced cars	approx. 14.450

Austin-Healey Sprite Mk I

The Austin-Healey Sprite debuted in 1958, a car whose price and performance lay somewhat below the "Big Healey". The small Sprite, which was soon given the nickname "frog" by its fans because of its unconventionally positioned headlights, developed into an affordable best-seller that additionally could be tuned relatively easily.

Model	Austin-Healey Sprite Mk I
Cubic capacity/cylinder	948 cc/4 cylinder
hp/kW	43/31
Construction period	1958 – 1961
Number of produced cars	approx. 39.000

Austin-Healey 3000 Mk III

Carefully improvement details contributed to the fact that the "Big Healey" lost nothing of its charm during later years. From 1964, with the launch of the 3000 Mk III version, it achieved its best form, which can be clearly indicated by looking at the sales figures. Today, cars of this model that could reach 190 km/h and featured wooden fascias are among the most popular of the whole series.

Model	Austin-Healey 3000 Mk III
Cubic capacity/cylinder	2912 cc/6 cylinder
hp/kW	150/110
Construction period	1964 – 1967
Number of produced cars	–

Bristol 405

The British aeroplane manufacturer Bristol had been considering building cars since the 1940s. When the first Bristol was launched in 1947, the similarities with BMW models were quite obvious. Bristol at first favoured BMW engine technology, before selecting American V8 engines. The Type 405 was incidentally the first sporty, four-door Bristol.

Model	Bristol 405
Cubic capacity/cylinder	1971 cc/6 cylinder
hp/kW	107/78.3
Construction period	1953 – 1957
Number of produced cars	approx. 300

Healey 2.4 Litre

Many years before the Austin-Healey was produced by the British Motor Corporation, Donald Healey already devoted himself to the construction of sporty cars. The first Healey, produced in a small series, was presented to the general public in 1946. Its body (convertible, roadster or coupé) was made of light alloy, and it was fitted with an engine made by Riley.

Model	Healey 2.4 Litre
Cubic capacity/cylinder	2443 cc/4 cylinder
hp/kW	106/78
Construction period	1946 – 1954
Number of produced cars	–

Healey Silverstone

The name of this flat racer is indicative of its function – a vehicle with the model name Silverstone could have only been conceived for racing. This is why Donald Healey opted for a lightweight mode of construction – in connection with a strong four-cylinder engine this allowed the racer to reach a maximum speed of approximately 180 km/h.

Model	Healey Silverstone
Cubic capacity/cylinder	2043 cc/4 cylinder
hp/kW	105/77
Construction period	1949 – 1951
Number of produced cars	105

Jaguar XK 120 Showcar

In 1948, Jaguar presented two concept cars at the London Motor show, the XK 100 and the XK 120. The resonance among visitors was overwhelming – while hardly anyone was interested in the four-cylinder XK 100, the company could have filled its order books for the XK 120 already at the fair. For Jaguar there was therefore only one right decision – this car had to go into production.

Model	Jaguar XK 120 Showcar
Cubic capacity/cylinder	3442 cc/6 cylinder
hp/kW	162/119
Construction period	1948
Number of produced cars	unique piece

Jaguar XK 120 Roadster

It would have been difficult to conceive of a more beautiful sports car in the 1940s. All its dimensions and proportions were right – including the long bonnet, short cockpit, low windshield and harmonious rear end. Added to this was the output potential of a strong six-cylinder engine with two overhead camshafts and exemplary road-holding.

Model	Jaguar XK 120 Roadster
Cubic capacity/cylinder	3442 cc/6 cylinder
hp/kW	162/119
Construction period	1948 – 1954
Number of produced cars	12.087

Jaguar XK 120 DHC

When Jaguar started the production of the new XK 120 in 1948, they initially only introduced one body variation, the roadster. In 1951, the coupé version was added, and in 1953 appeared the "Drop Head Coupé". While a Roadster disposed only of a light canvas soft top, this version was equipped with a thoroughly lined convertible top and crank handle windows in the doors.

Model	Jaguar XK 120 DHC
Cubic capacity/cylinder	3442 cc/6 cylinder
hp/kW	213/156
Construction period	1948 – 1954
Number of produced cars	12.087

Jaguar XK 140

In the course of model development, Jaguar introduced the XK 140 in 1954. In addition to technical modifications (more output) there were also many modifications to the car's appearance. These suited the XK excellently, because they corresponded to the preferences of the time – the body silhouette was somewhat tautened, the radiator grille slightly enlarged, the bumpers strengthened, and the interior refined.

Model	Jaguar XK 140
Cubic capacity/cylinder	3442 cc/6 cylinder
hp/kW	192/140
Construction period	1954 – 1957
Number of produced cars	8884

Jaguar XK 140 Coupé

What did the model abbreviation "XK" actually stand for? The answer to this justifiable question is rather simple – "X" stood for "experimental", and "K" originated from a series of internal codes for Jaguar's engine development. Of course, the fact that the project "XK" should one day become the foundation of a sports car myth could not be anticipated by anyone at the time.

Model	Jaguar XK 140 Coupé
Cubic capacity/cylinder	3442 cc/6 cylinder
hp/kW	192/140
Construction period	1954 – 1957
Number of produced cars	8884

Jaguar XK 150

All models of the XK class contributed to British automotive history and in particular the final series – the XK 150 – contributed to the fame of the brand. In addition, it also constituted the basis for future models. In line with the common practice of the era, a great number of accessories were available for the XK 150, including for example, wire-spoked wheels equipped with whitewall tyres or an overdrive.

Model	Jaguar XK 150
Cubic capacity/cylinder	3442 cc/6 cylinder
hp/kW	213/156
Construction period	1957–1961
Number of produced cars	9395

Jaguar XK 150

Following the big success of the XK class, the company attempted to also set up the XK in the USA. The model thrilled many Americans, and they were particularly enthusiastic about the about the last type, the XK 150, which was once again an even further development in terms of looks and technology. While coupés and convertibles were produced from 1957 to 1961, the roadster was only made between 1958 and 1960.

Model	Jaguar XK 150
Cubic capacity/cylinder	3442 cc/6 cylinder
hp/kW	213/156
Construction period	1957 – 1961
Number of produced cars	9395

Jaguar E-Type Series 1

When in 1961 an aesthetically stylish two-seater called Jaguar E-Type debuted in Geneva, nobody could have anticipated that this car should initiate a new chapter in British sports car history. Under the long bonnet of the E-Type was a 265 hp six-cylinder engine. It accelerated the car weighing 1168 kilogram to 240 km/h – while allowing it to reach the 100 km/h – mark after only seven seconds.

Model	Jaguar E-Type Series 1
Cubic capacity/cylinder	3781 cc/6 cylinder
hp/kW	265/196
Construction period	1961–1964
Number of produced cars	15 700

Jaguar E-Type Series 2

In addition to the convertible version, Jaguar introduced the elegant E-Type also as "2+2-seater" coupé. While the appearance of this variation with the boot lid opening to the left took some getting used to, it also offered its own particular type of charm. Due to its fast-back body, the wheel base was extended, providing the passengers seated on the jump seat back bench with a bit more head room.

Model	Jaguar E-Type Series 2
Cubic capacity/cylinder	4235 cc/6 cylinder
hp/kW	265/194
Construction period	1968–1971
Number of produced cars	18.820

Jaguar E-Type Series 2

The Jaguar E-Type was certainly among the cars, which the general public was excited about from the day of their first-time presentation. Its justified position as an "icon of automotive construction" was documented in 1996 also by the "Museum of Modern Art" in New York, that included the convertible version of the E-Type as one of only three significant cars in its permanent exhibition.

Model	Jaguar E-Type Series 2
Cubic capacity/cylinder	4235 cc/6 cylinder
hp/kW	265/194
Construction period	1968 – 1971
Number of produced cars	18.820

Jaguar XJ 13

Jaguar actually intended the XJ 13 to be a race car model. However, before the car had a chance to become involved in racing, it fell victim to the Group policy. Nevertheless, the effort of its development was not in vain – Jaguar engineers finally derived from the beefy V12 engine an engine for the serially-produced cars aimed at clients with racing ambitions.

Model	Jaguar XJ 13
Cubic capacity/cylinder	4994 cc/12 cylinder
hp/kW	500/368
Construction period	1966
Number of produced cars	unique piece

Lotus Seven Serie 1

In 1947, Colin Chapman, founder of the British brand Lotus, constructed a small sports car in his spare time. The success of this model motivated him ten years later to convert his amateur garage into a factory and to begin serial production. Because in England fewer taxes were levied on kit cars, Chapman offered the "Lotus Seven" primarily as a kit car.

Model	Lotus Seven Serie 1
Cubic capacity/cylinder	1172 cc/4 cylinder
hp/kW	40/29
Construction period	1957 – 1970
Number of produced cars	–

Lotus Elite

The Lotus Cars Ltd founded in 1955 did not need long to become internationally renown. The sporty vehicles developed by Colin Chapman were enthusiastically received from the start with their light synthetic bodywork and an engine that could be very easily tuned.

Model	Lotus Elite
Cubic capacity/cylinder	1216 cc/4 cylinder
hp/kW	76/56
Construction period	1957 – 1963
Number of produced cars	988

Lotus Elan S1

A little more comfort – this was the essential innovation which distinguished the Elan from its predecessor, the Elite. In addition, as opposed to the Elite, the Elan was subjected to regular model development (S2, S3, S4 and Sprint). The 1.5-litre engine positioned in the front transmitted its power (between 106 hp and 126 hp) to the back wheels, which provided the car with a top speed between 185 km/h and 195 km/h.

Model	Lotus Elan S1
Cubic capacity/cylinder	1558 cc/4 cylinder
hp/kW	126/93
Construction period	1962 – 1973
Number of produced cars	12 224

Lotus Europe

In 1966, Lotus introduced the Europe model, a ultra-flat, mid-engine sports car that was only 113 centimetres high. The automotive press was quite astonished when it discovered the engine unit under the hood: it consisted of the engine of a Renault R 16. Because the Lotus was very light thanks to its synthetic bodywork, a top speed of 200 km/h could be achieved with this configuration.

Model	Lotus Europe
Cubic capacity/cylinder	1470 cc/4 cylinder
hp/kW	78/57
Construction period	1967 – 1975
Number of produced cars	9230

MG TC

The small TC, introduced by the British brand MG to its program after the end of the Second World War, was a vehicle which soon helped the company to regain its excellent market position. The fact that the TC was based on an outdated box frame hardly bothered anybody. After all, other manufacturers also continued their pre-war constructions after resuming production.

Model	MG TC
Cubic capacity/cylinder	1250 cc/4 cylinder
hp/kW	54/40
Construction period	1945 – 1949
Number of produced cars	approx. 10.000

MG TC

With its clearly structured technology, the MG TC immediately embodied the perfect example of a typically British sports car. It was not only its affordable price that made it interesting to many fans. Within a short period, the best seller also conquered the hearts of American automobile enthusiasts. Another pleasant quality of the TC was the fact that it could be tuned easily.

Model	MG TC
Cubic capacity/cylinder	1250 cc/4 cylinder
hp/kW	54/40
Construction period	1945–1949
Number of produced cars	approx. 10.000

MG TD

When MG introduced the TD model as the successor of the TC in 1949, only very close inspection revealed the improvement of details. Overall, the TD gained some width, its cockpit was not laid out quite as narrowly as before. Unfortunately, the large wire-spoked wheels had to give way to conventional steel rims. One advantage of this measure was that the TD looked less long-legged than the TC.

Model	MG TD
Cubic capacity/cylinder	1250 cc/4 cylinder
hp/kW	55/40
Construction period	1949 – 1953
Number of produced cars	approx. 30.000

MG TF

While almost all manufacturers introduced smoothly styled bodies in the 1950s, MG continued the design of the 1930s with its models TC, TD and TF. They noticed too late that this style was not in demand anymore, and the sales plunged drastically. While the face-lift for the revised Type TF was well intentioned, it turned out to be rather ineffective in the end.

Model	MG TF
Cubic capacity/cylinder	1250 cc/4 cylinder
hp/kW	54/40
Construction period	1953 – 1954
Number of produced cars	approx. 6200

MG TF 1500

To make up for its lost business, MG launched the TF in an even more improved version. A 1.5-litre engine allowed the TF 1500 to actually achieve a speed of 140 km/h. However, this measure could not positively affect the sales success anymore – the time of the T series, of which about 50.000 units across models had been produced, had definitively come to an end.

Model	MG TF 1500
Cubic capacity/cylinder	1466 cc/4 cylinder
hp/kW	64/47
Construction period	1954 – 1955
Number of produced cars	approx. 3400

MG A

In 1955, the MG A was a sports car that was launched at the exactly right time to help the MG brand regain some of its lost popularity. Softly curved lines dominated the design of the two-seater that differed completely from the TC, TD or TF. The MG A was based on a foundation that allowed the majority of this model to be constructed as a roadster.

Model	MG A
Cubic capacity/cylinder	1489 cc/4 cylinder
hp/kW	69/51
Construction period	1955 – 1962
Number of produced cars	approx. 90.000

MG A Coupé

One could not drive an open MG A coupé. The roof was firmly connected with the body, which of course had the advantage that the coupé was a weatherproof vehicle. Compared with the roadster, the coupés were more luxuriously equipped, with an upholstered fascia, fitted carpet and triangular vent windows.

Model	MG A Coupé
Cubic capacity/cylinder	1489 cc/4 cylinder
hp/kW	69/51
Construction period	1955 – 1962
Number of produced cars	approx. 98.900

MG A Twin-Cam

At the launch of the car, testers already agreed that the MG A could have used a stronger engine. The fact that MG took this suggestion up in 1958 and immediately offered a model with a twin-cam shaft engine, was a pleasant surprise for the fans. Now the MG A Twin-Cam disposed of the road performance of a Porsche 356 Carrera – for much less money!

Model	MG A Twin-Cam
Cubic capacity/cylinder	1589 cc/4 cylinder
hp/kW	108/79
Construction period	1958 – 1960
Number of produced cars	2111

MG A Twin-Cam

Without doubt, the MGA Twin-Cam was one of the most beautiful British sports cars. Unfortunately, it also had a few negative qualities – its complex engine with two overhead camshafts required a large degree of instinct in everyday handling. The engine disliked humid weather as well as downhill driving. In addition, the oil consumption increased substantially with a sporty manner of driving.

Model	MG A Twin-Cam
Cubic capacity/cylinder	1589 cc/4 cylinder
hp/kW	108/79
Construction period	1958 – 1960
Number of produced cars	2111

Morgan 4/4

The British company Morgan initially became famous for its sporty "Threewheelers", before they built in 1936 their first "regular" four-wheeled car. The basic principle of this rock-hard suspension roadster remained largely unchanged until the late 1960s. Anyone who would like to purchase a handmade Morgan must be very patient – the delivery time is quite long.

Model	Morgan 4/4
Cubic capacity/cylinder	1498 cc/4 cylinder
hp/kW	64/47
Construction period	1962 – 1967
Number of produced cars	–

Morgan + 4

Once something was at one point proven to be useful, Morgan considered changing it a complete waste of time. Therefore, the company has been building this sports car model almost without any changes for decades. The most substantial remodelling took place in 1955 when the cooler was changed from flat to round. However, under its hood, Morgan is very contemporary, featuring state-of-the-art technology.

Model	Morgan + 4
Cubic capacity/cylinder	2138 cc/4 cylinder
hp/kW	105/77
Construction period	1950 – 1958
Number of produced cars	–

Morgan Plus 8

In 1968, Morgan drivers were presented with a totally new driving option – since that was the year the new "Plus 8" became available. Its name derived from the fact that it was equipped with the 3.5-litre V8 engine by Rover. With the help of this engine, the speedometer needle reached the vicinity of the 200-km/h mark. To allow the light Morgan to handle this much power, amendments included expanding the track width to 126 centimetres.

Model	Morgan Plus 8
Cubic capacity/cylinder	3532 cc/8 cylinder
hp/kW	184/135
Construction period	1968 – 2004
Number of produced cars	approx. 6000

Sunbeam Alpine

In addition to Hillman, Humber and Singer, the Rootes Group also included the brand Sunbeam. Sunbeam introduced the Alpine, an open sports car with a rather more gentle character than the MG or Triumph models. The Alpine was rarely found in races, as the large two-seater convertible with its quality interior belonged on the regular street.

Model	Sunbeam Alpine
Cubic capacity/cylinder	1725 cc/4 cylinder
hp/kW	94/69
Construction period	1959 – 1961
Number of produced cars	–

Sunbeam Alpine II

Until early 1961, the bodies for the Alpine model were produced by the former Armstrong Siddeley works. Beginning in summer 1961 however, Sunbeam commissioned Pressed Steel Ltd in Oxford with the construction of the coaches. The Alpine cars made after the production site change were immediately recognisable by their radiator grilles with horizontally arranged ornamental bars.

Model	Sunbeam Alpine II
Cubic capacity/cylinder	1592 cc/4 cylinder
hp/kW	86/63
Construction period	1961 – 1963
Number of produced cars	–

Triumph TR 2

In 1952, Triumph exhibited at the London Motor Show a roadster with very low-cut doors, extended fender skirts and the opening of the radiator grille moved far to the inside. The resonance to the TR 2 was overwhelming. One year later already, the 170 km/h-fast model went into production. To be able to reach 170 km/h, the engine was equipped, incidentally, with two horizontal SU carburettors.

Model	Triumph TR 2
Cubic capacity/cylinder	1991 cc/4 cylinder
hp/kW	91/67
Construction period	1953 – 1955
Number of produced cars	–

Triumph TR 3 A

In 1955, in the course of model development, Triumph juxtaposed the stronger model TR 3 A to the TR 2. In line with the technical progress, the TR 3 A was equipped with disc brakes in the front. In addition, the model featured a newly designed and more softly upholstered fascia, broader seats and, above all, a smart radiator grille that ran across the whole width of the car.

Model	Triumph TR 3 A
Cubic capacity/cylinder	1991 cc/4 cylinder
hp/kW	101/74
Construction period	1957 – 1961
Number of produced cars	–

Triumph TR 4

In search of a successor for the TR 3, Triumph soon faced a problem: to be able to build a modern car they had to replace the outdated chassis. Giovanni Michelotti, an experienced chassis designer, was commissioned to find a solution. Under his direction, the very modern TR 4 was created the first sports car in Great Britain with a fully synchronised transmission.

Model	Triumph TR 4
Cubic capacity/cylinder	1991 cc/4 cylinder
hp/kW	100/73
Construction period	1961 – 1967
Number of produced cars	71.665

Triumph TR 6

There was a reason why the TR 6, introduced in 1969, resembled the TR 4 – it also featured a design by Michelotti, which was however, reworked once more before the beginning of the serial production by Karmann in Osnabrück. In contrast to the TR 4, the TR 6 was equipped with a six-cylinder rather than a four-cylinder engine. Based on its rather high output of 143 hp, the TR 6 achieved a maximum speed of 200 km/h.

Model	Triumph TR 6
Cubic capacity/cylinder	2498 cc/6 cylinder
hp/kW	143/105
Construction period	1969 – 1976
Number of produced cars	94.619

TVR Grantura

In 1947, young Trevor Wilkinson began to produce his first sports car. He used the chassis of the Alvis Firebird for the car's undercarriage – however, only two years later the gifted Briton started to develop a chassis of his own. From the outset, TVR vehicles gained many victories in British races, giving the company an excellent reputation.

Model	TVR Grantura
Cubic capacity/cylinder	1798 cc/4 cylinder
hp/kW	98/72
Construction period	1958 – 1967
Number of produced cars	–

Alpine A 110

When the son of a French Renault trader built a sports car in 1958 on the basis of 4 CV, hardly anyone was interested. By 1960 however, the specialist press already knew his cars, and in 1963 he achieved the great breakthrough with the Alpine A 110 model – which initiated serial production. Depending on each car's achievement potential, the Alpine A 110 could reach high speeds of up to 215 km/h.

Model	Alpine A 110
Cubic capacity/cylinder	1565 cc/4 cylinder
hp/kW	140/103
Construction period	1963 – 1976
Number of produced cars	7160

Bugatti 101 C

The Bugatti 101 C, which celebrated its debut in 1951 in Paris, continued the glorious legacy of the enterprise. Coach maker Gangloff in Colmar manufactured its contemporary ponton body. Under its bonnet an eight-cylinder engine was at work that accelerated the 101 C to a speed of 180 km/h. The models equipped with a compressor received the name 101 C.

Model	Bugatti 101 C
Cubic capacity/cylinder	3257 cc/8 cylinder
hp/kW	190/139
Construction period	1952 – 1954
Number of produced cars	7

Delahaye 175 S

Magnificent car bodies could be mounted on the chassis of the Delahaye 175 S, which the French noble brand starting building in 1948. After all, the undercarriage had a wheel base of nearly 300 centimetres. To enable comfortable driving of the imposing 175 S, it was equipped with a semiautomatic four-speed transmission that was electrically controlled by a tiny lever on the fascia.

Model	Delahaye 175 S
Cubic capacity/cylinder	4455 cc/6 cylinder
hp/kW	140/103
Construction period	1948 – 1949
Number of produced cars	–

DB Le Mans

Starting in the 1950s, Charles Deutsch and René Bonnet, two French engineers, introduced various sports cars to the market under the brand name "DB". Their models were for the most part powered by a two-cylinder engine (!) by Panhard. The little engines were often tuned up to the outer limits of their capacity, because this was the only way to achieve top speeds of 150 km/h or more.

Model	DB Le Mans
Cubic capacity/cylinder	848 cc/2 cylinder
hp/kW	52/38
Construction period	1960 – 1962
Number of produced cars	approx. 200

Matra Djet V

In 1964, Matra released a sports car with a medium-sized engine that was comparable to any deluxe vehicle of its time not in terms of technology, but definitely in terms of interior design. The eye-catcher of the little car was a wide middle console loaded with a multitude of round instruments. The undercarriage of the "Djet V", a light space frame, was constructed in-house while the engine was supplied by Renault.

Model	Matra Djet V
Cubic capacity/cylinder	1255 cc/4 cylinder
hp/kW	72/53
Construction period	1964 – 1968
Number of produced cars	1681

Matra M 530

In 1967, Matra introduced the model M 530 to the market as the successor of the Djet. The bodywork of this rather strange looking car consisted of plastic synthetics and it was powered by a Ford engine. As a result of the ongoing concerns about security at the time, Matra equipped the nearly 200 km/h-fast M 530 with disc brakes on all four wheels.

Model	Matra M 530
Cubic capacity/cylinder	1699 cc/4 cylinder
hp/kW	70/51
Construction period	1967 – 1973
Number of produced cars	9609

Peugeot 504 Coupé

The 504 by Peugeot was a contemporarily-styled car with a slightly sporty touch. The two-door model was offered as a convertible and as a coupé. Automobile designer Pininfarina designed the contours of the two-door car. As opposed to the four-door 504 sedan, the coupé and convertible versions were based on a foundation with a shortened wheel base (255 centimetres).

Model	Peugeot 504 Coupé
Cubic capacity/cylinder	2664 cc/6 cylinder
hp/kW	144/105.5
Construction period	1969 – 1983
Number of produced cars	26.000

Peugeot 504 Cabriolet

Until 1974, the 504 was equipped with a four-cylinder engine (1971 cubic centimetres, 100 hp). All subsequent model versions were equipped with a powerful V6 engine. Judging by the number of produced cars, the customers obviously favoured the elegant coupé variation. It was produced 26,000 times, as opposed to a mere 8135 convertible units.

Model	Peugeot 504 Cabriolet
Cubic capacity/cylinder	2664 cc/6 cylinder
hp/kW	144/105.5
Construction period	1969–1983
Number of produced cars	8135

Alfa Romeo 6 C 2500

The first Alfa Romeo produced after the end of the Second World War strongly resembled the models of the 1930s. Luxurious models like this 6 C 2500 were still impressive, but definitely no longer modern. To be able to survive on the market, Alfa Romeo restructured the plant in 1950 to meet the requirements of modern mass production.

Model	Alfa Romeo 6 C 2500
Cubic capacity/cylinder	2443 cc/6 cylinder
hp/kW	90/66
Construction period	1946 – 1950
Number of produced cars	–

Alfa Romeo 2500 Villa d'Este

In the late 1940s, Alfa Romeo and Pininfarina continued to expand their joint avant-garde reputation. The results of their efforts were repeatedly awarded with prizes, for example at the famous "Concorso d'Eleganza Villa D'Este" in Lake Como. In 1949, they presented a car named after the event, the Alfa Romeo 2500 Villa d'Este.

Model	Alfa Romeo 2500 Villa D'Este
Cubic capacity/cylinder	2443 cc/6 cylinder
hp/kW	90/66
Construction period	1949
Number of produced cars	–

Alfa Romeo 1900 Super Sprint

Soon after Alfa Romeo presented the sporty 1900 Sprint, rumours began to circulate that the quick coupé would soon become available in a stronger version. And the insiders were right – in 1954 the 115 hp Super Sprint was released. This 180 km/h-fast two-door model with its long wheel base was distinguished by an elegance rarely found in comparable models.

Model	Alfa Romeo 1900 Super Sprint
Cubic capacity/cylinder	1975 cc/4 cylinder
hp/kW	115/84
Construction period	1954 – 1958
Number of produced cars	–

Alfa Romeo 1900 Sprint

The Alfa Romeo 1900 was a four-door sedan which divers with racing ambitions had been eagerly awaiting for a long time. The four-cylinder engine featured two overhead camshafts as well as a cylinder head of light alloy. This technology also inspired the coach maker Touring who subsequently manufactured a sporty two-seater, the "Sprint", based on the sedan.

Model	Alfa Romeo 1900 Sprint
Cubic capacity/cylinder	1975 cc/4 cylinder
hp/kW	90/66
Construction period	1953 – 1958
Number of produced cars	–

Alfa Romeo 1900 Super Sprint

To really get the Alfa Romeo 1900 Super Sprint going, the first series of the car was equipped with a 100 hp, 1.9-litre four-cylinder engine. This engine provided the coupé, which weighed approximately 1050 kilograms, with a speed of around 180 km/h. In 1956, with the debut of the model 1900 SS (Super Sprint), the output was increased to 115 hp. The car's price tag of DM 31,000 also matched its classy characteristics.

Model	Alfa Romeo 1900 Super Sprint
Cubic capacity/cylinder	1975 cc/4 cylinder
hp/kW	115/84
Construction period	1954 – 1958
Number of produced cars	–

Alfa Romeo Disco Volante

While private drivers with racing ambitions proved their talent in the 1950s with their Alfa Romeos on the racetracks, the manufacturer produced some racing sports cars models based on the street vehicles. One of these was the Spider that was derived from the 1900, with a distinctive bodywork that spontaneously provided it with the nickname "Disco Volante" (i.e. "Flying saucer").

Model	Alfa Romeo Disco Volante
Cubic capacity/cylinder	1997 cc/4 cylinder
hp/kW	158/116
Construction period	1952
Number of produced cars	6

Alfa Romeo Giulietta Sprint

The 1.3-litre and unexpectedly light aluminium four-cylinder engine with which the first Giulietta sprint series was equipped featured a cross-flow cylinder head and two overhead camshafts – at that time this corresponded to the latest in racing technology. The result of the development efforts was an initial 65 hp, enough for a maximum speed of about 165 km/h.

Model	Alfa Romeo Giulietta Sprint
Cubic capacity/cylinder	1290 cc/4 cylinder
hp/kW	65/48
Construction period	1954 – 1964
Number of produced cars	approx. 36.000

Alfa Romeo Giulietta Sprint Speciale

In response to the constant demand for better performance, the horsepower of the Giulietta was steadily increased in the course of model development – instead of 65 hp, soon the rear wheels were powered by 80 hp, and even 90 hp in the Sprint Veloce. The sporty variations Sprint Zagato (SZ) and sprint Speciale (SS) even benefited from 100 hp, which allowed the lightweight automobile to easily reach the 200-km/h mark.

Model	Alfa Romeo Giulietta Sprint Speciale
Cubic capacity/cylinder	1570 cc/4 cylinder
hp/kW	100/74
Construction period	1957 – 1965
Number of produced cars	approx. 2760

Alfa Romeo Giulietta Spider

For many Giulietta owners, the most intriguing property of the car was its technology. It had an engine with two overhead camshafts, independent suspension and a well-gradated manual transmission. Two years after the debut of the Giulietta followed the next models in its evolution: for racing this consisted of the Giulietta Sprint Veloce, and for private drivers it was the Giulietta Spider.

Model	Alfa Romeo Giulietta Spider
Cubic capacity/cylinder	1290 cc/4 cylinder
hp/kW	65/48
Construction period	1955 – 1965
Number of produced cars	approx. 26.400

Alfa Romeo Giulietta Spider

Because the Alfa Romeo Spider, a spin-off of the sedan model "Sprint", enjoyed growing popularity, the manufacturer complied faster than expected with a customer wish and increased the cubic capacity to 1.6 litres. The fact that the serially provided collapsible hood rendered the Spider only partially weatherproof did not diminish its sales success. After all, there was also an optional hard top available.

Model	Alfa Romeo Giulietta Spider
Cubic capacity/cylinder	1570 cc/4 cylinder
hp/kW	112/82
Construction period	1955 – 1965
Number of produced cars	approx. 26.346

Alfa Romeo Giulia Sprint GT

1963 witnessed the birth of an automotive history classic – the breathtakingly beautiful Giulia Sprint GT. The successor of the Giulia Sprint was also sometimes known as the "step front bonnet" because of its initially stacked-up looking bonnet. The no-frills bodywork design was by Bertone, where it actually was the first design by the new chief designer Giorgetto Giugiaro.

Model	Alfa Romeo Giulia Sprint GT
Cubic capacity/cylinder	1290 cc/4 cylinder
hp/kW	80/59
Construction period	1963 – 1968
Number of produced cars	approx. 222.000

Alfa Romeo Giulia Sprint GTA

Giorgetto Giugiaro, today's head of Italdesign, designed the Giulia Sprint GT, a car that already evolved into a legend during its construction period. It was offered in numerous variations on the market and was widely acknowledged in racing as Giulia Sprint GTA. Incidentally, the GTA won no less than seven EM titles!

Model	Alfa Romeo Giulia Sprint GTA
Cubic capacity/cylinder	1570 cc/4 cylinder
hp/kW	115/01
Construction period	1965 – 1970
Number of produced cars	–

Alfa Romeo Canguro

In 1964, Nuccio Bertone created together with Giorgetto Guigiaro the Canguro coupé, which was shown as a prototype at the Mondial de L'Automobile (Paris Motor Show). Technically, this show car was based on the Alfa Romeo TZ 1. The aerodynamically shape of the Canguro gave the four-cylinder car a relatively "light" and quick look: 200 km/h were easily achieved.

Model	Alfa Romeo Canguro
Cubic capacity/cylinder	1570 cc/4 cylinder
hp/kW	113/83
Construction period	1964
Number of produced cars	unique piece

Alfa Romeo Giulia TZ 2

When a customer in 1956 brought his Giulietta Sprint GT that had been involved in accident, to the coachmaker Zagato for repairs, it was the starting signal for a new sports car generation. In 1963, in the course of its development, this Giulietta SZ mutated into the stronger Giulia TZ (Tubulare Zagato) which regularly caused sensations in racing.

Model	Alfa Romeo Giulia TZ 2
Cubic capacity/cylinder	1570 cc/4 cylinder
hp/kW	170/125
Construction period	1963 – 1967
Number of produced cars	–

Alfa Romeo Iguana

It is a pity that the Alfa Romeo Iguana created by Italdesign did not go into production – with this just 100 centimetres-high vehicle, Alfa Romeo would have been offering a competitive car to the Maserati. This rear-wheel drive coupé with a mid-engine (L x W x H = 4050 x 1780 x 1050 millimetres) displayed excellent road holding and was equipped with a short-stroke V8 engine.

Model	Alfa Romeo Iguana
Cubic capacity/cylinder	1995 cc/8 cylinder
hp/kW	233/171
Construction period	1969
Number of produced cars	unique piece

Alfa Romeo Duetto

When Alfa Romeo in 1965 discontinued the Giulietta Spider, the most successful convertible of the time, the following model was ready to be launched. While it took some time for the "Duetto", the successor whose production commenced in 1966, to be accepted, in good time this model also proved to be very successful and thrilled many convertible drivers for nearly three decades.

Model	Alfa Romeo Duetto
Cubic capacity/cylinder	1570 cc/4 cylinder
hp/kW	92/67
Construction period	1966 – 1982
Number of produced cars	–

Alfa Romeo Spider

The Alfa Romeo Duetto, successor of the Giulietta-Spider, experienced a difficult start, because its rear was not generally accepted. Hence, in 1970 it was redesigned and provided with a trailing edge. Increasing sales figures led to an expansion of the model range – in addition to the Spider 1300 junior version it also included the Spider 1600, 1750 and 2000.

Model	Alfa Romeo Spider
Cubic capacity/cylinder	1570 cc/4 cylinder
hp/kW	92/67
Construction period	1966 – 1982
Number of produced cars	–

Bizzarrini GT Strada 5300

Before Giotto Bizzarrini started his own business in 1961, the gifted Italian had been working for Alfa Romeo and Enzo Ferrari, among others. In 1965 he introduced his GT Strada 5300, a visually attractive sports car under whose aluminium body a large-volume Chevrolet engine was at work bringing the car to a top speed of 270 km/h.

Model	Bizzarrini GT Strada 5300
Cubic capacity/cylinder	5351 cc/8 cylinder
hp/kW	350/257
Construction period	1965 – 1969
Number of produced cars	149

Bizzarrini Manta

Bizzarrini's model range did not offer much variety. In addition to the 5300 GT, he also launched the model 1900 Europe, a smaller sports car whose production was stopped in 1969 however, after producing only 15 vehicles. The Bizzarrini Manta shown here did not go into production at all. The only 1050 millimetre-high coupé, by the way, was a design by Giugiaro.

Model	Bizzarrini Manta
Cubic capacity/cylinder	5351 cc/8 cylinder
hp/kW	350/257
Construction period	1969
Number of produced cars	unique piece

Dino 246 GT

In 1967, Ferrari launched under the autonomous brand Dino a small mid-engine sports car, the Dino 206 GT of which about 150 units were built until 1969. In the second edition – the Dino 246 GT – the car sold considerably better. Incidentally, the automotive press considered the 246 GT to be a competitor of Porsche. Which car customers preferred to the other was entirely a matter of taste.

Model	Dino 246 GT
Cubic capacity/cylinder	2418 cc/6 cylinder
hp/kW	190/139
Construction period	1969 – 1974
Number of produced cars	3883

Ferrari 342 America

In racing the name Ferrari was firmly established since the 1930s, because Enzo Ferrari achieved one victory after the other. However, the first street sports car (Type 166) carrying his name did not debut until in 1948. Of course, it was its classy twelve-cylinder engine that made this car so desirable. From that point on, this engine served as the base for further car models.

Model	Ferrari 342 Americana
Cubic capacity/cylinder	4102 cc/12 cylinder
hp/kW	200/147
Construction period	1952 – 1953
Number of produced cars	6

Ferrari 375 America

Almost each early Ferrari was an individual unique piece, because Enzo Ferrari co-operated simultaneously with several car body specialists. As a result, of the 375 America, built a mere 12 times, eight bodies originated from Pininfarina, three from Vignale and one was constructed by Ghia. This much individualism had its price – most of these cars were ordered by celebrities.

Model	Ferrari 375 Americana
Cubic capacity/cylinder	4523 cc/12 cylinder
hp/kW	300/220
Construction period	1953 – 1955
Number of produced cars	12

Ferrari 410 Spyder Superamerica

In 1956, Ferrari introduced the successor of the 375 America, the even substantially stronger "410 Superamerica". To be able to draw ample strength from this model of the five-litre class, the beefy V12 aggregate was equipped with three downdraft double carburettors. Depending on the rear-axle transmission, the maximum speed was up to 265 km/h – for a 1950s car a more than satisfactory value.

Model	Ferrari 410 Spyder Superamerica
Cubic capacity/cylinder	4963 cc/12 cylinder
hp/kW	340/249
Construction period	1955
Number of produced cars	2

Ferrari 410 Superamerica

For three years, the 410 Superamerica was featured in the model range of Ferrari. During this time, only 37 (!) units were actually produced – automobile construction cannot get more individual. Of these 37 cars which each had a different look, 33 were "outfitted" by the studio of bodywork construction expert Pininfarina, while the remaining four were designed by his competitors.

Model	Ferrari 410 Superamerica
Cubic capacity/cylinder	4963 cc/12 cylinder
hp/kW	340/249
Construction period	1956 – 1959
Number of produced cars	37

Ferrari 250 GT SWB

For Ferrari fans, the abbreviation SWB holds a very special meaning – it is derived from the English term short wheel base. For the Ferrari 250 GT SWB, the wheel base amounts to exactly 240 centimetres, and based upon this pleasant characteristic, the car could not only be driven excellently on the street, but on racetracks as well.

Model	Ferrari 250 GT SWB
Cubic capacity/cylinder	2953 cc/12 cylinder
hp/kW	260/190
Construction period	1959 – 1962
Number of produced cars	165

Ferrari 250 GT SWB

The American magazine "Sports Car Illustrated" was given in 1960 the pleasure to be allowed to test the 250 GT SWB. Shortly afterwards it wrote that "The latest Ferrari Berlinetta radiates the essence of speed and strength, ... This is a fast car, very potent, almost brutal, ... Without exaggeration it can be said that it is powered by the greatest engine of our times."

Model	Ferrari 250 GT SWB
Cubic capacity/cylinder	2953 cc/12 cylinder
hp/kW	260/190
Construction period	1959 – 1962
Number of produced cars	165

Ferrari 250 GT „Tour de France"

Ferrari commissioned Pininfarina to build four prototypes with different appearances, before giving the "green light" for the construction of the 250 GT. Nevertheless, the design of the serial models was constantly changing and – differing from the usual custom – the racing cars of the 250 GT series very much resembled the street cars. Incidentally, the cockpit of 250 GT was narrow and noisy, yet this did in no way diminish the joy of driving this car.

Model	Ferrari 250 GT
Cubic capacity/cylinder	2953 cc/12 cylinder
hp/kW	270/198
Construction period	1956 – 1959
Number of produced cars	84

Ferrari 250 GT Cabriolet

Their general appearance puts Ferraris among the cars one enjoys staring after. Small wonder, since in most cases the design of the cars originated on the drawing board of Pininfarina, the Italian master coach maker. He also designed the silhouette of the 250 GT convertible. No easy task, since this model was based on a chassis with a wheel base of 260 centimetres.

Model	Ferrari 250 GT Cabriolet
Cubic capacity/cylinder	2953 cc/12 cylinder
hp/kW	240/176
Construction period	1957 – 1962
Number of produced cars	236

Ferrari 250 GT Spyder California

The 250 GT Spyder California could not be described more suitably than by the "Sports Car Illustrated" magazine: "The California has the most beautiful (coach) body this side of the Riviera. We do not know how or why, but the Italians seem to own an exclusive contract for automotive beauty. In short, we consider the bodywork, the engine and the transmission to be terrific ...".

Model	Ferrari 250 GT Spyder California
Cubic capacity/cylinder	2953 cc/12 cylinder
hp/kW	280/205
Construction period	1957 – 1963
Number of produced cars	104

Ferrari 250 GTO

Although it was road certified, the Ferrari 250 GTO was most comfortable on the racetrack. Maybe this was due to the fact that the engineers had had a look at the Testarossa during its construction. Here also, the engine was placed deep within the tubular frame. Because the GTO was conceived as a competitive car from the start, Pininfarina gave it a particularly streamlined body.

Model	Ferrari 250 GTO
Cubic capacity/cylinder	2953 cc/12 cylinder
hp/kW	300/220
Construction period	1962 – 1964
Number of produced cars	36

Ferrari 275 GTB

The elegantly shaped Berlinetta 275 GTB, which Ferrari presented in 1964, was more than a stylish coupé for everyday use. This Ferrari was at the same time a demanding sports car for active drivers – there were even some "tuned up" 275 GTBs that participated successfully in the Targa Florio or in Le Mans.

Model	Ferrari 250 GTB
Cubic capacity/cylinder	3286 cc/12 cylinder
hp/kW	280/205
Construction period	1964 – 1966
Number of produced cars	472

Ferrari 275 GTB–N.A.R.T. Spider

Ferrari built the 275 GTB also as a Spider version. However, this stylistically completely different edition did not come close to looking as attractive as the Berlinetta. The N.A.R.T. Spider, however, was an exception – with approval of the manufacturer it was made in the USA by the Ferrari importer Luigi Chinetti. The abbreviation N.A.R.T. in the model name stands for "North America Racing Team".

Model	Ferrari 275 GTB–N.A.R.T. Spider
Cubic capacity/cylinder	3286 cc/12 cylinder
hp/kW	300/220
Construction period	1966 – 1968
Number of produced cars	10

Ferrari 330 GT 2+2

With the launch of the 330 GT 2+2, Ferrari for the first time introduced a car which, in the opinion of the automotive press, was unfortunately a little bit too ungainly – everyone had become accustomed to more thrilling cars by Ferrari. Ferrari had to subject to the criticism and modified the looks at the launch of the second series. The 330 GT 2+2 was hardly suited for racing but so for longdistance journeys.

Model	Ferrari 330 GT 2+2
Cubic capacity/cylinder	3967 cc/12 cylinder
hp/kW	300/220
Construction period	1964–1967
Number of produced cars	1085

Ferrari 330 GTC

The US magazine "Car and Driver" in 1967 provided its readers with the following advice for the 330 GTC: "Step on the clutch, turn the ignition key and give the gas pedal a light, nervous touch." Who did this, got to experience that the 330 GTC was not a well-behaved little coupé, but an aggressive sports device whose speedometer needle did not come to a halt before 242 km/h.

Model	Ferrari 330 GTC
Cubic capacity/cylinder	3967 cc/12 cylinder
hp/kW	300/220
Construction period	1966–1968
Number of produced cars	604

Ferrari 365 GTB/4

As common for Ferrari models, the number 365 provided information about the cubic capacity of one cylinder. Accordingly, the 365 GTB/4 had an overall cubic capacity of 4.4 litres. The "4" in the model name referred to the four overhead camshafts of the engine. The 365 GTB/4 also known as the "Ferrari Daytona", reached the then breathtaking top speed of 275 km/h.

Model	Ferrari 365 GTB/4
Cubic capacity/cylinder	4390 cc/12 cylinder
hp/kW	352/258
Construction period	1968–1973
Number of produced cars	1245

Ferrari 365 GTS/4

With the 365 GTB/4, Ferrari undoubtedly introduced one of the most beautiful sports cars of all times. However, its front part had to be changed shortly after the start of serial production because the American licensing laws prescribed another position for the flood-lights. As of 1969, the 365 GBT/4 was also available in a Spider version that was released under the model name 365 GTS/4.

Model	Ferrari 365 GTS/4
Cubic capacity/cylinder	4390 cc/12 cylinder
hp/kW	352/258
Construction period	1969 – 1973
Number of produced cars	121

Fiat Abarth 850 TC

The brand emblem featuring a scorpion is well-known in the world of competitive racing. Wherever it appears, one immediately knows that Carlo Abarth has once again transformed a simple standard car into something sporty. The Fiat 850 TC Abarth was also created under his direction. Believe it or not, this fiery vehicle is actually based on the little Fiat 600.

Model	Fiat Abarth 850 TC
Cubic capacity/cylinder	847 cc/4 cylinder
hp/kW	62/45
Construction period	1956 – 1964
Number of produced cars	–

Fiat 1200 Spider TV

Actually, the Fiat 1200 Spider TV was intended to be a very special car. However, the mixture of Italian design and American silhouette did not sell well at all. The rear part of the car was obviously modelled on the Lancia Aurelia Spider. If the panoramic windshield had been a little bigger, it could have absolutely originated from Cadillac's spare parts.

Model	Fiat 1200 Spider TV
Cubic capacity/cylinder	1221 cc/4 cylinder
hp/kW	55/40
Construction period	1955 – 1959
Number of produced cars	3400

Fiat 1200 Stanguellini Spider

Starting in 1946, Vittorio Stanguellini in Modena focused on refining Fiat cars. He was particularly specialised in engine modifications, while every now and then he also fitted Fiat chassis with elegant special coaches. Some of his ideas were implemented as small series. Others – like the Spider shown here – were created only as exclusive unique piece.

Model	Fiat 1200 Stanguellini Spider
Cubic capacity/cylinder	1221 cc/4 cylinder
hp/kW	55/40
Construction period	1957
Number of produced cars	unique piece

Fiat 2300 S Coupé

In spite of its size, the Fiat 2300 S Coupé was no family car. In the back there was only one auxiliary bench which did not even comfortably accommodate children. However, in terms of looks, the car absolutely matched the preferences of the time, because Ghia in designing the body included a tripartite rear window as a special style element. The six-cylinder engine powered the coupé providing a maximum speed of 190 km/h.

Model	Fiat 2300 S Coupé
Cubic capacity/cylinder	2279 cc/6 cylinder
hp/kW	136/100
Construction period	1961 – 1968
Number of produced cars	–

Fiat 124 Sport Spider

In the 1960s, there was probably no other automobile manufacturer with a model range as extensive as that of Fiat. Fiat offered everything: from the smallest car to deluxe vehicles – and, of course, the popular 124 Spider. Those who were enthusiastic about this model made a good choice – the Fiat 124 cost an equivalent of EUR 1000 less than the competitive model by Alfa Romeo.

Model	Fiat 124 Sport Spider
Cubic capacity/cylinder	1438 cc/4 cylinder
hp/kW	90/66
Construction period	1966 – 1982
Number of produced cars	approx. 130.000

Fiat 124 Sport Coupé

At the launch of its career, the Fiat 124 had to be content with a 1.4-litre engine. Later, the cubic capacity was expanded to 1.6 and 1.8 litres respectively, while the last construction series (180 km/h high speed) even benefited from a two-litre engine. Many Fiat 124 cars were exported to the USA, and to fulfil the ardent exhaust laws there they were equipped with an injection system.

Model	Fiat 124 Sport Coupé
Cubic capacity/cylinder	1995 cc/4 cylinder
hp/kW	118/86
Construction period	1966 – 1982
Number of produced cars	approx. 130.000

Fiat Dino Spider

Critics often claimed that the Dino Spider was nothing more than a cheap Ferrari. This was absolutely wrong, because this model was a one hundred percent indigenous Fiat construction. Only the engine was a Ferrari design. This spirited V6 engine (initially with two-litres and later 2.4 litres of cubic capacity) with the overhead camshafts enabled the Spider to reach a top speed of 210 km/h.

Model	Fiat Dino Spider
Cubic capacity/cylinder	1987 cc/6 cylinder
hp/kW	160/117
Construction period	1966 – 1972
Number of produced cars	approx. 1580

Fiat Dino Coupé

The Dino Coupé, which was based in contrast to the Spider on a long wheel base (255 instead of 228 centimetres), practically asked to be driven forcefully. It accelerated excellently and became especially spirited in the upper speed range. Bertone designed its bodywork, while star designer Pininfarina was responsible for the design of the Spider version.

Model	Fiat Dino Coupé
Cubic capacity/cylinder	2418 cc/6 cylinder
hp/kW	180/132
Construction period	1967 – 1972
Number of produced cars	approx. 4200

Iso Rivolta IR 300

Long before Iso specialised in the construction of sports cars, the head of the company Renzo Rivolta had established himself in the automotive branch – he had developed the small car construction, which was later produced by BMW as the "BMW Isetta". With another construction – the large Iso Rivolta IR 300 – he ventured in 1962 to enter into the sports car business himself.

Model	Iso Rivolta IR 300
Cubic capacity/cylinder	5354 cc/8 cylinder
hp/kW	304/224
Construction period	1961 – 1965
[illegible]	[illegible]

Iso Grifo GL 365

After the rather ill fated start of the Iso Rivolta in 1961, Renzo Rivolta launched another model two years later: the Iso Grifo. Bertone designed the cars that were equipped with a V8 Chevrolet engine (with a cubic capacity ranging according to model between 5.4 and 7 litres). He provided the massive coupé with a fairly harmonious silhouette.

Model	Iso Grifo GL 365
Cubic capacity/cylinder	5354 cc/8 cylinder
hp/kW	365/267
Construction period	1965 – 1966
Number of produced cars	–

Iso Grifo GL 365

Renzo Rivolta had only one purpose in mind when launching the Grifo – his cars should instil fear in Enzo Ferrari. Although Isos reached maximum speeds of up to 270 km/h according to their power transmission, these cars simply lacked the personality and flair of Ferraris. Following Rivolta's death, the decline of the brand was already imminent, the last car was manufactured in 1974.

Model	Iso Grifo GL 365
Cubic capacity/cylinder	5354 cc/8 cylinder
hp/kW	365/267
Construction period	1965 – 1966
Number of produced cars	–

Lamborghini 350 GTV

The aura of Lamborghini is connected to the man who made his own dreams come true: As a child, Ferruccio Lamborghini already showed interest in technology and mechanics. In 1959, he began to build tractors in his own company, and with this financial background he soon opened a car factory, where the prototype of the first Lamborghini shown here originated in 1963.

Model	Lamborghini 350 GTV
Cubic capacity/cylinder	3497 cc/12 cylinder
hp/kW	360/264
Construction period	1963
Number of produced cars	2

Lamborghini 350 GT

The 350 GT, a further development of the 350 GTV prototype, was equipped with a beefy V12 engine with four overhead camshafts. The power of the engine was transmitted to the rear axle by a five-speed gearbox. Because the resonance to the pre-serial model was rather positive, Lamborghini in 1963 already gave the "green light" for the production of the first serial model.

Model	Lamborghini 350 GT
Cubic capacity/cylinder	3646 cc/12 cylinder
hp/kW	270/198
Construction period	1963 – 1966
Number of produced cars	143

Lamborghini 350 GTS

The legend began in late 1963 in Sant'Agata, at the company headquarters of Automobili Lamborghini. With short intervals, the company developed a wide model range, which corresponded to the prevailing paragon of beauty and perfection. As an alternative to the 2 + 2-seat coupé version of the 350 GT, an open Spyder was also available at the traders for test-drives.

Model	Lamborghini 350 GTS
Cubic capacity/cylinder	3467 cc/12 cylinder
hp/kW	270/198
Construction period	1964
Number of produced cars	2

Lamborghini 400 GT 2 + 2

Until 1972, Lamborghini's company grew steadily, the only factor hampering the growth being paradoxically the introduction of new models, which slowed down the production capacity every now and then. During the manufacturing period of the 350 GT already, a version with an increased output, the 400 GT 2 + 2, was introduced. In addition to technical modifications, this car benefited from an exceedingly successful face-lift.

Model	Lamborghini 400 GT 2 + 2
Cubic capacity/cylinder	3929 cc/12 cylinder
hp/kW	320/234
Construction period	[illegible]
Number of produced cars	247

Lamborghini Miura P 400

Designed by Lamborghini's chief designer Marcello Gandini, the Miura P 400 with a horizontally fitted four-litre V12 engine celebrated its launch in 1966 at the Geneva International Motor Show. With this model, the Italian noble brand presented for the first time a classy mid-engine sports car, that strengthened the reputation of the enterprise as the maker of spectacular cars.

Model	Lamborghini Miura P 400
Cubic capacity/cylinder	3929 cc/12 cylinder
hp/kW	320/234
Construction period	1966 – 1969
Number of produced cars	475

Lamborghini Miura P 400 S

For the Miura P 400 SV, the second revised version of the Miura P 400 prototype of 1966, the front and back wheel suspension were completely redesigned, the tyre equipment adjusted and the wings made a little more prominent. These changes in the new model resulted in an athletic-looking sports car whose four litre V12 engine had an output of approximately 370 hp.

Model	Lamborghini Miura P 400 S
Cubic capacity/cylinder	3929 cc/12 cylinder
hp/kW	370/271
Construction period	1969 – 1971
Number of produced cars	140

Lamborghini Miura Spider

The low roofline was the most fascinating feature of the Miura. The elegance of the only 105 centimetre-high racecar is considered an automotive paradigm of classic design. This could have even been improved on – in 1968, Lamborghini showed a Miura Spider at the Brussels International Motor Show. Sports car fans fervently hoped that this show car would be produced as a serial model. Unfortunately, their hope was in vain.

Model	Lamborghini Miura Spider
Cubic capacity/cylinder	3929 cc/12 cylinder
hp/kW	320/234
Construction period	1968
Number of produced cars	unique piece

Lamborghini Islero

The Islero, which complemented the model range as of 1968, also had a V12 engine with four camshafts, independent suspension and disc brakes. This car was primarily conceived for businessmen who wanted to travel fast and sportily. Although a little shorter than 400 GT 2 + 2, the Islero, which was manufactured in different performance ranges, offered a generously designed interior.

Model	Lamborghini Islero
Cubic capacity/cylinder	3929 cc/12 cylinder
hp/kW	340/249
Construction period	1968 – 1969
Number of produced cars	225

Lamborghini Espada

In 1968 debuted the exceptionally stylish Espada model. The exterior of the car with a top speed of 250 km/h strongly resembled the Marzal, a concept car presented by Bertone the previous year. The extremely low Espada combined the looks, performance and handling of a top sports car with the comfort and the luxury of a sedan. Against all expectations, the Espada became one of the most successful Lamborghinis ever.

Model	Lamborghini Espada
Cubic capacity/cylinder	3929 cc/12 cylinder
hp/kW	350/256
Construction period	1968 – 1978
Number of produced cars	1217

Lancia Aurelia GT B 20

Lancia's model GT B 20 – an elegant fastback coupé – was intended as an alternative for individuals in search of a thrilling four-door sedan. While the rear bench of this two-door model was very cramped, the driver and front seat passenger sat in veritable leather armchairs. The fascia was dominated by three big round instruments, which the driver could see very well.

Model	Lancia Aurelia GT B 20
Cubic capacity/cylinder	1991 cc/6 cylinder
hp/kW	75/55
Construction period	1951 – 1953
Number of produced cars	–

Lancia Aurelia GT B 20

Because the GT B 20 performed excellently not only on the street but also on the racetrack, the first tuning specialists soon detected a worthwhile business opportunity and offered all kinds of the accessories with which the engine power could be increased in a sustainable way. The manufacturer also regularly increased the performance – following the first "hp boost" the GT B 20 already achieved about 185 km/h.

Model	Lancia Aurelia GT B 20
Cubic capacity/cylinder	1991 cc/6 cylinder
hp/kW	75/55
Construction period	1951 – 1953
Number of produced cars	–

Lancia Aurelia B 24 Spider

Immediately following its launch, the automotive press was sure that with the Aurelia Spider, Lancia had launched one of the most beautiful and most fascinating sports cars of the 1950s. After it had built 240 cars, in the course of model development some optical and technical improvements were made, and the correctly criticised convertible soft-top construction was improved as well.

Model	Lancia Aurelia B 24 Spider
Cubic capacity/cylinder	2458 cc/6 cylinder
hp/kW	108/79
Construction period	[illegible]
Number of produced cars	–

Lancia Aurelia B 24 Spider

At the beginning of the year 1954, the second generation of the Aurelia class was available at the retailers for test runs. The most important constructive innovation was a completely revised rear axle, which had been derived from the racing sports car D 24. Over the course of time engine modifications were regularly implemented, which made the car increasingly more dynamic: its achievement performance increased to 118 hp.

Model	Lancia Aurelia B 24 Spider
Cubic capacity/cylinder	2458 cc/6 cylinder
hp/kW	108/79
Construction period	1956 – 1959
Number of produced cars	–

Lancia Fulvia Sport

For Lancia, the era of the front-wheel drive began in 1960 with the middle class sedan Flavia. Its chassis layout was also applied to the compact Fulvia Coupé. A good decision, which immediately advanced the Fulvia, of which there was also a rare sports version with a Zagato body, to the best-seller of the Lancia porgramme.

Model	Lancia Fulvia Sport
Cubic capacity/cylinder	1584 cc/4 cylinder
hp/kW	114/84
Construction period	1965 – 1972
Number of produced cars	approx. 6000

Maserati A6 GCS

After Maserati had built many top-class racing sports cars in the 1930s, in 1946 the enterprise constructed for the first time a street sports car, the A6. The A6, which was based on a well thought-out plan from the outset, remained for a long time the basic pillar of the model range. Its cubic capacity was regularly increased and its power output boosted.

Model	Maserati A6 GCS
Cubic capacity/cylinder	1985 cc/6 cylinder
hp/kW	167/122
Construction period	1953 – 1957
Number of produced cars	–

Maserati 3500 GT

In 1957, Maserati presented a surprising two-seater coupé that was powered by an engine with two overhead camshafts. In addition, the 6-cylinder engine featured a double ignition and was served by three double carburettors. The body of the 230 km/h-fast coupé was produced as a lightweight construction.

Model	Maserati 3500 GT
Cubic capacity/cylinder	3485 cc/6 cylinder
hp/kW	220/161
Construction period	1958 – 1964
Number of produced cars	approx. 2000

Maserati 3500 GT Spider

In 1947, after the movement of the company headquarters to Modena, Maserati for the first time focused more intensely on the construction of street sports cars. As expected, the first model was already received enthusiastically, and when the company in 1958 presented the 3500 GT, the brand had been established in the sports car sector for a long time. The body of the 3500 was produced as a lightweight construction.

Model	Maserati 3500 GT Spider
Cubic capacity/cylinder	3485 cc/6 cylinder
hp/kW	220/161
Construction period	1958 – 1964
Number of produced cars	approx. 2000

Maserati 5000 GT

In search of even more performance, Maserati developed in the late 1950s a sports car in the five-litre class, the 5000 GT. It was equipped with a V8 engine and was only made to order. Coach maker Touring provided most of the bodies of these 270 km/h-fast cars built in a lightweight construction style developed by him (System Superleggera).

Model	Maserati 5000 GT
Cubic capacity/cylinder	4975 cc/8 cylinder
hp/kW	350/256
Construction period	1959 – 1965
Number of produced cars	–

Maserati Sebring

Strictly speaking, the Sebring was an especially sporty variation of the 3500 GT, but with a frame that was shortened by about ten centimetres. To be able to save weight, the steel sheet body constructed by Vignale in Turin was given a hood of light alloy. In the course of model development the power output was varied several times, and accordingly the car's top speed ranged between 235 km/h and 245 km/h.

Model	Maserati Sebring
Cubic capacity/cylinder	3485 cc/6 cylinder
hp/kW	255/188
Construction period	1961 – 1969
Number of produced cars	446

Maserati Mistral Spider

It was a common practice at Maserati to name the fastest model of the brand after a type of wind. This practice began with the Mistral in 1963. Its streamlined silhouette was based on a design of the coach maker Pietro Frua. The Mistral – which came in the model versions coupé and spider – was built for seven years and was accordingly often subjected to model development.

Model	Maserati Mistral Spider
Cubic capacity/cylinder	3693 cc/6 cylinder
hp/kW	245/180
Construction period	1963 – 1970
Number of produced cars	approx. 120

Maserati Indy

While the competition had been equipping their cars for a long time already with rear independent suspensions, in the late 1960s the Maserati Indy remained content with a rigid rear axis and leaf springs. The according to factory specifications fully fledged four-seater, was equipped with a V8 engine and achieved a top speed of 245 km/h – in 1969, other four-seaters could not beat this value.

Model	Maserati Indy
Cubic capacity/cylinder	4136 cc/8 cylinder
hp/kW	260/190
Construction period	1968 – 1974
Number of produced cars	1136

Siata 750 Sports Spider

It was definitely possible to transform a simple Fiat into a sharp sports car, if the job was carried out by a specialised expert, for instance the Siata company. In 1952, it introduced to the market a Barchetta that had been equipped with an American Crossley engine. This engine with overhead camshaft accelerated the little Siata to a top speed of 140 km/h.

Model	Siata 750 Sports Spider
Cubic capacity/cylinder	721 cc/4 cylinder
hp/kW	32/23
Construction period	1952 – 1954
Number of produced cars	approx. 40

Saab Sonett I

Riveted aluminium sheets are normally used for aeroplane construction. However, they can also serve, as in case of the Saab Sonett, as the foundation for a car. Saab, which was familiar with the aviation industry, opted for this unusual material and as a result the base of its concept car "Sonett I" was much more rigid and at the same time lighter than the conventional steel constructions.

Model	Saab Sonett I
Cubic capacity/cylinder	748 cc/3 cylinder
hp/kW	58/42
Construction period	1956
Number of produced cars	unique piece

Saab Sonett I

While the foundation of the Sonett I was made of aluminium, its body consisted of light GRP material (plastic reinforced by glass-fibre). As a result, the little sports car only topped the scales at 500 kilograms. Its top speed was 160 km/h. As opposed to the original plans, Saab produced the Sonett in a small series of six units.

Model	Saab Sonett I
Cubic capacity/cylinder	748 cc/3 cylinder
hp/kW	56/41
Construction period	1956
Number of produced cars	6

Saab Sonett II

Finally, after several unsuccessful approaches, Saab commenced with the production of a little sports car with a synthetic body in 1966. Developed based on an experimental vehicle, the car with a wheel base of only 215 centimetres was called Sonett II and equipped with a tuned three-cylinder engine. For stability reasons, the construction received additional steel stabiliser bars.

Model	Saab Sonett II
Cubic capacity/cylinder	841 cc/3 cylinder
hp/kW	60/44
Construction period	1966 – 1970
Number of produced cars	[illegible]

Saab Sonett II

In the opinion of many sports car fans, equipping the Saab Sonett II with a three-cylinder two-stroke engine was actually a disgrace. Saab had to realise this fact this and soon used a four-cylinder engine by Ford. Although the maximum speed was now 160 km/h, nevertheless, the expected increase in sales figures did not follow.

Model	Saab Sonett II
Cubic capacity/cylinder	1498 cc/4 cylinder
hp/kW	65/48
Construction period	1966 – 1970
Number of produced cars	1498

Volvo P 1900 Sport

It is rather unusual for a manufacturer of trucks and respectable sedans to suddenly present a sports car. This is exactly what Volvo did in 1954. Surprisingly, the bodywork was made of glass fibre. Because hardly anyone would buy such a car in the cold north, the P 1900 Sport was primarily intended for the export market.

Model	Volvo P 1900 Sport
Cubic capacity/cylinder	1414 cc/4 cylinder
hp/kW	70/51
Construction period	1956 – 1957
Number of produced cars	67

Volvo P 1900 Sport

In the 1950s in general, cars with synthetic bodies were initially received with scepticism by buyers. This problem was well known to Chevrolet (Corvette) as well as Volvo. It was not easy to successfully position a "plastic car" in the market. Volvo probably had to struggle with this problem for too long, which is why the production of the interesting P 1900 Sport was finally discontinued in 1957.

Model	Volvo P 1900 Sport
Cubic capacity/cylinder	1414 cc/4 cylinder
hp/kW	70/51
Construction period	1956 – 1957
Number of produced cars	67

Monteverdi 375 S

Peter Monteverdi was not only an experienced racing driver, but also a gifted design engineer who implemented his dream of a sports car made by him in 1967. Pietro Frua supported the Swiss engineer with this plan and created the coachwork for the car. The Monteverdi was a luxurious car that was powered by an American V8 engine.

Model	Monteverdi 375 S
Cubic capacity/cylinder	7206 cc/8 cylinder
hp/kW	375/275
Construction period	1967 – 1969
Number of produced cars	–

Buick Wildcat I

As of the year 1949 – the year the first in-house General Motors exhibition called "Motorama" was established – it became standard procedure for the GM Group to present show cars in addition to serial models. The individual brands of the Group presented any vehicle that in their view could become a reality in the future. In 1953, Buick's exhibition contribution included the concept car Wildcat I.

Model	Buick Wildcat I
Cubic capacity/cylinder	5276 cc/8 cylinder
hp/kW	188/138
Construction period	1953
Number of produced cars	individual piece

Buick Wildcat II

Following the brilliant success of the Wildcat I (1953) and under the direction of Harley Earl, the head of design of the GM Group, Buick developed for the 1955 Motorama the Wildcat II. This model, which was styled according to the early retro look, appeared quite aggressive by its exterior alone. The show car was equipped with modern synthetic bodywork, as was the later built Wildcat III.

Model	Buick Wildcat II
Cubic capacity/cylinder	5276 cc/8 cylinder
hp/kW	202/148
Construction period	1955
Number of produced cars	2

Chevrolet Corvette

The main attraction of the 1953 Motorama was of course the presentation of the new Corvette. Several versions of this sports car with the fibre glass body were presented simultaneously, including a practical station wagon as well as an elegant coupé with gracefully suggested rear fins. Despite the beauty of all these experiments, in the end only the roadster was implemented.

Model	Chevrolet Corvette
Cubic capacity/cylinder	3859 cc/6 cylinder
hp/kW	150/110
Construction period	1953
Number of produced cars	various show cars

Chevrolet Corvette

For many years now the name Corvette has been firmly established in the sports car scene. However, this was not always the case, because the very first Corvette had a problem – its engine power and road performance did not live up to expectations. However, with intensive model development this situation was quickly remedied and the car, which was initially equipped with a plastic body, has today attained the rank of a cult object.

Model	Chevrolet Corvette
Cubic capacity/cylinder	3859 cc/6 cylinder
hp/kW	150/110
Construction period	1953 – 1955
Number of produced cars	4640

Chevrolet Corvette

At the beginning of its career, the Corvette was only available in white. In contrast, the interior of the two-seater was designed completely in red. In addition, the silhouette of the plastic bodywork was accentuated with chrome moulding. However, as opposed to the "Rear-fin Yanks" Chevrolet rather held back with the use of chrome.

Model	Chevrolet Corvette
Cubic capacity/cylinder	3859 cc/6 cylinder
hp/kW	150/110
Construction period	1953 – 1955
Number of produced cars	4640

Chevrolet Corvette

The attractive design and the chrome-plated stone shields above the headlights gave the Corvette a distinctively sporty look. However, instead of the expected eight-cylinder engine, it was initially only equipped with a six-cylinder engine. It was only under the direction of design engineer Zora Arkus-Duntov that the Corvette developed into an eight-cylinder dream car.

Model	Chevrolet Corvette
Cubic capacity/cylinder	4342 cc/8 cylinder
hp/kW	195/143
Construction period	1956 – 1962
Number of produced cars	64.375

Chevrolet Corvette

When Chevrolets head design engineer Duntov equipped the Corvette in 1955 with a V8 engine, it was possible for the first time to order a manual three-gear transmission instead of the automatic gear. In 1956, the designers decided on a comprehensive face-lift: The optical retouching suited the sports car nicely, and in 1958, with the introduction of double headlights, the Corvette looked better than ever before.

Model	Chevrolet Corvette
Cubic capacity/cylinder	4342 cc/8 cylinder
hp/kW	195/143
Construction period	1956 – 1962
Number of produced cars	64.375

Chevrolet Corvette

When Ford presented the Thunderbird in 1954, it looked as if the Corvette had received a competitor. However, it is common knowledge that the Thunderbird developed a short time later into a four-seater, and Chevrolet could once again assume its former position. In 1954 incidentally, the Corvette profited from an important technical innovation – its outdated six-volt electrical system was replaced by a modern twelve-volt arrangement.

Model	Chevrolet Corvette
Cubic capacity/cylinder	4342 cc/8 cylinder
hp/kW	225/165
Construction period	1956 – 1962
Number of produced cars	64.375

Chevrolet Corvette Sting Ray

The new outfit of the Corvette in which it appeared as of 1963 had no longer been designed under the responsibility of Harley Earl, since at that point designer Bill Mitchell already shaped the appearance of the car. Mitchell first of all designed the so-called split window, a divided rear window that only existed in 1963. The technology also had something new to offer – finally the Corvette had an independent rear suspension.

Model	Chevrolet Corvette Sting Ray
Cubic capacity/cylinder	5359 cc/8 cylinder
hp/kW	250/183
Construction period	1963 – 1967
Number of produced cars	45.546 (coupés only)

Chevrolet Corvette Sting Ray Cabriolet

In addition to the closed-body Sting Ray with the "split-window", Corvette enthusiasts consider the Sting Ray Convertible to be among the most successful achievements of Bill Mitchell. As of 1965, the Sting Ray finally received front disc braking, instead of the four-wheel drum brakes. In addition, this model was the first Corvette equipped with retractable headlights.

Model	Chevrolet Corvette Sting Ray Cabriolet
Cubic capacity/cylinder	5359 cc/8 cylinder
hp/kW	250/183
Construction period	1963 – 1967
Number of produced cars	–

Chevrolet Corvette Sting Ray

In 1967, already the fourth generation of Corvettes left the conveyor belts, and, as once noted by a journalist, was already teeming with power while simply sitting there. Or at least while its 7.4-litre V8 engine was rumbling under its bonnet. By the way, owners of a closed-body Corvette could remove the upper section of the roof and transform their coupé with a few moves into a "fresh air car".

Model	Chevrolet Corvette Sting Ray
Cubic capacity/cylinder	7440 cc/8 cylinder
hp/kW	465/341
Construction period	1967 – 1974
Number of produced cars	approx. 150.000

Chevrolet Corvette Stingray

In the course of model development, the fourth Corvette generation received the probably most radical design overhaul of the car's history. The opulent shapes were of course also applied to the convertible version, incidentally. As opposed to the earlier models, as of 1969 the name "Stingray" was spelled as a single word and no longer as two words ("Sting Ray").

Model	Chevrolet Corvette Stingray
Cubic capacity/cylinder	7440 cc/8 cylinder
hp/kW	465/341
Construction period	1967 – 1974
Number of produced cars	–

Dodge Charger

In the European market, the Dodge Charger definitely belonged to the obscure Greats. This racer, whose headlights were skilfully integrated into the radiator grille, was serially equipped with a 5.2-litre V8 engine. Who wanted to go beyond the 200 km/h mark, was forced to order the seven-litre engine, the so-called "Street Hemi", at a surcharge.

Model	Dodge Charger
Cubic capacity/cylinder	5210 cc/8 cylinder
hp/kW	233/171
Construction period	1966 – 1967
Number of produced cars	–

Dodge Charger

The Dodge Charger proved to be true to its powerful name. With just under three meters of wheel base, it was definitely not among the most manageable American cars, but decidedly among the most potent ones. Although its silhouette was less attractive than that of its predecessors, the new shape became popular in its time, since it included the fashionable "hip-swing".

Model	Dodge Charger
Cubic capacity/cylinder	6286 cc/8 cylinder
hp/kW	274/210
Construction period	1967 – 1969
Number of produced cars	–

Dodge Charger

In terms of sales figures, the Charger was the secondbest-selling vehicle of Dodge. Reason enough to improve it further in the course of model development. Depending on the engine configuration, the maximum speed of the extremely heavy car was between 160 km/h and 230 km/h – in the late 1960s, petrol consumption still played a subordinate role.

Model	Dodge Charger
Cubic capacity/cylinder	7206 cc/8 cylinder
hp/kW	380/279
Construction period	1968 – 1970
Number of produced cars	–

Dodge Charger Coronet Super Bee

In 1969, a war over horsepower raved among the US car manufacturers. Sportier, stronger and faster was the slogan of the day. This horsepower rumpus rarely resulted in quality sports cars, with the trend being in favour of muscle cars. Dodge was diligently involved in the action – here for example is a husky street racer called "Super Bee".

Model	Dodge Charger Coronet Super Bee
Cubic capacity/cylinder	7195 cc/8 cylinder
hp/kW	390/287
Construction period	1969 – 1970
Number of produced cars	–

Dodge Charger III Concept Car

With the Charger III concept car Dodge only wanted to test the reaction of the general public and a serial production of the two-seater sports cars was not intended at the time. The Charger III was rather used to display new style elements and technologies. Whether intentional or not – the Charger III was the most aerodynamic car Dodge ever created.

Model	Dodge Charger III
Cubic capacity/cylinder	no information available
hp/kW	no information available
Construction period	1968
Number of produced cars	concept car/ unique specimen

Ford Thunderbird

When the first orders for the Thunderbird started to come in, it was expected that the annual production would be approximately 10,000 – however, the retailers presented the Group with 16,000 orders. In 1957/58, a total of 21,000 orders filled the order books, and there was no end in sight for this upward trend. It seemed that everybody wanted to own this "real car", which had a steel body in contrast to the Corvette.

Model	Ford Thunderbird
Cubic capacity/cylinder	5113 cc/8 cylinder
hp/kW	210/154
Construction period	1955 – 1957
Number of produced cars	53.166

Ford Thunderbird

While the GM Group was already working on its Corvette project in 1951, Ford had also been thinking about a similar car. However, the idea was quickly shelved – too quick, it turned out. When Chevrolet's Corvette was already receiving plenty of attention in 1953, competitor Ford had to make up for the lost time, which is why the Thunderbird only became available for purchase two years later.

Model	Ford Thunderbird
Cubic capacity/cylinder	4780 cc/8 cylinder
hp/kW	193/141
Construction period	1955 – 1957
Number of produced cars	53.166

Ford Thunderbird

Quite deliberately the Ford Group gave its first sports car the model name Thunderbird, because among native Americans the thunderbird has been considered a token of good luck for aeons. The bird was believed to bring power and prosperity, and that was exactly what Ford needed at that time. It should be noted that to the media, the Thunderbird was a "Sports car with American Luxury" – its design was considered too European.

Model	Ford Thunderbird
Cubic capacity/cylinder	4780 cc/8 cylinder
hp/kW	193/141
Construction period	1955 – 1957
Number of produced cars	53.166

Ford Thunderbird

Apparently, Thunderbird customers were among the drivers who preferred the rather comfortable driving style – this could be derived from the fact that Ford delivered the majority of all its models with the "Fordomatic" option. This was an interesting alternative to the manual three-gear transmission and made driving the car with the hefty V8 engine a true pleasure.

Model	Ford Thunderbird
Cubic capacity/cylinder	5113 cc/8 cylinder
hp/kW	210/154
Construction period	1955 – 1957
Number of produced cars	53.166

Ford Thunderbird

The fact that over the course of the time the size of the Thunderbird engine was continuously increased was definitely a pleasant development. Unfortunately, the proportions of the body also grew every year, and the former very elegantly styled two-seater slowly mutated into a street cruiser. Starting with the second series, the Thunderbird that had previously only been built as a convertible also became available as a coupé.

Model	Ford Thunderbird
Cubic capacity/cylinder	5766 cc/8 cylinder
hp/kW	304/223
Construction period	1958 – 1960
Number of produced cars	approx. 200.000

Ford Thunderbird Series III

In the opinion of the designers, the Thunderbird was in urgent need of a new look for the model year 1961. To the shocked amazement of many fans they were faced with a car with a flatly stretched out body – Ford described it quite elegantly as a "cigar shape". Be that as it may, the T-Bird was available with many output levels and as usual as a hardtop coupé, convertible and roadster.

Model	Ford Thunderbird Series III
Cubic capacity/cylinder	6348 cc/8 cylinder
hp/kW	304/223
Construction period	1961 – 1963
Number of produced cars	approx. 214.375

Ford Mustang

If US citizens were in search of a slightly smaller car in the 1960s, they were forced to revert to European import models. Lee Iacocca, who headed Ford at the time, wanted to change this. His answer to an Austin Healey or a Fiat Dino was called a "Mustang". This model was available for purchase as of 1964.

Model	Ford Mustang
Cubic capacity/cylinder	3273 cc/6 cylinder
hp/kW	122/89
Construction period	1964 – 1967
Number of produced cars	–

Ford Mustang

Ford was lucky enough to receive in excess of 22.000 orders immediately upon the debut of the Mustang. While all these early models were equipped with a six-cylinder engine, the expansion of the model range by adding models with hefty V8 engines was soon to follow. In addition, as of 1966, Ford equipped the Mustang with front disc brakes.

Model	Ford Mustang
Cubic capacity/cylinder	4728 cc/8 cylinder
hp/kW	228/167
Construction period	1964 – 1967
Number of produced cars	–

Ford Mustang

Within the first production year of the Mustang, a total of 680.000 units had already been produced, and an end of the boom was not in sight. The sporty trend setting car cost about USD 2300 in its most basic version, but with a long list of accessories. Broader tyres, a brake power assist unit, a revolution counter and seat belts were among the most popular extras.

Model	Ford Mustang
Cubic capacity/cylinder	4728 cc/8 cylinder
hp/kW	275/201
Construction period	1964 – 1967
Number of produced cars	–

Ford Mustang

With the Mustang, Ford targeted affluent young buyers who did not want to park a virtuous old sedan in their garage. Because Ford built the Mustang from the start as a large-scale production, it could offer the model at an especially favourable price. When the first Mustangs were finally available, the customers literally snapped them out of the hands of the traders – the equation had worked out.

Model	Ford Mustang
Cubic capacity/cylinder	4728 cc/8 cylinder
hp/kW	203/149
Construction period	1964 – 1967
Number of produced cars	–

Ford Mustang

In 1966 appeared the last Mustang with a triple taillight that could be immediately assigned to the first series. The fact that Ford had in the meantime also introduced a fastback coupé model did not in any way change the fact that the convertible style was the absolute favourite among buyers.

Model	Ford Mustang
Cubic capacity/cylinder	4728 cc/8 cylinder
hp/kW	275/201
Construction period	1964 – 1967
Number of produced cars	–

Ford Mustang Shelby GT 500

In 1967, the proportions of the Mustang started to expand, and the former beautifully designed car increased in length and width. Carroll Shelby, who was specialised in the tuning of Mustang models, was hardly bothered by this. His customers wanted only one thing – output. Shelby's creation for 1969 was called GT 500. If requested, a seven-litre V8 engine powered this broad and powerful machine.

Model	Ford Mustang Shelby GT 500
Cubic capacity/cylinder	7033 cc/8 cylinder
hp/kW	340/249
Construction period	1969 – 1970
Number of produced cars	–

Ford Mustang Shelby GT 500 KR

When in 1965 racer specialist Carroll Shelby tuned his first Mustang-models, he started by launching the GT 350 – a fastback coupé with 4.7 litres of cubic capacity. Because he refined the V8 engine with a special carburettor, racing intake manifold and a double exhaust system, suddenly 330 hp became available. An even substantially stronger model, the GT 500 KR, soon followed the 350 GT.

Model	Ford Mustang Shelby GT 500 KR
Cubic capacity/cylinder	6989 cc/8 cylinder
hp/kW	360/264
Construction period	1968 – 1970
Number of produced cars	–

Oldsmobile Golden Rocket

In 1956, Oldsmobile presented the Golden Rock model with a particularly unconventional silhouette – corresponding to how the engineers envisioned the streamlined design of the future. To dramatically present the concept car on the occasion of the Motorama show, the show car was heaved on the stage in front of the audience.

Model	Oldsmobile Golden Rocket
Cubic capacity/cylinder	5400 cc/8 cylinder
hp/kW	230/169
Construction period	1956
Number of produced cars	unique piece

Ford GT 40

In the 1960s, the Ford Group created the GT 40, a sports car, which today achieves record prices at vintage car auctions. The extremely low-slung two-seater was primarily developed for competitive racing. Based on the subsequent great demand, a limited street version was developed later.

Model	Ford GT 40
Cubic capacity/cylinder	4728 cc/8 cylinder
hp/kW	340/250
Construction period	1966 – 1972
Number of produced cars	107

Pontiac Firebird

The General Motors brand Pontiac added the Firebird to its programme range in the late 1960s. Its coachwork consisted of a unibody construction, which at the time was far from being the standard for all Pontiac cars. The Firebird was offered in several output classes, and disc brakes for the stronger models could be ordered at a surcharge.

Model	Pontiac Firebird
Cubic capacity/cylinder	5340 cc/8 cylinder
hp/kW	253/185
Construction period	1967 – 1969
Number of produced cars	–

Pontiac Bonneville

Not every car constructed by the American automobile manufacturers in the 1950s was equipped with massive rear fins. The Pontiac Bonneville, of which two units were built in 1954, presented a surprisingly different silhouette. The interior of the two-seater (made of red synthetic leather) was also worth looking at, while its fascia featured a long row of round instruments.

Model	Pontiac Bonneville
Cubic capacity/cylinder	4728 cc/8 cylinder
hp/kW	188/138
Construction period	1954
Number of produced cars	2

Holden Hurricane

In 1948, the first cars of the Holden brand rolled off the production lines at Fishermen's Bend in Australia. Initially an assembly plant of the General Motors Group, the company eventually also launched models of its own, including sports cars. Unfortunately, the futuristic Hurricane, a mid-engine coupé which, instead of doors, featured a canopy that swung open forward over the front wheels, remained in the prototype stage and never went into production.

Model	Holden Hurricane
Cubic capacity/cylinder	4141 cc/8 cylinder
hp/kW	243/179
Construction period	1969
Number of produced cars	unique piece

Honda S 800 Cabrio

Honda S 800 was one of the first cars that the Japanese conglomerate exported successfully to Europe. Compared with its predecessor, the S 600, the power of the S 800 was not longer transmitted to the rear axle by a chain, but rather by means of a hypoid gear. In addition, the little sports car was equipped with front disc brakes.

Model	Honda S 800 Cabrio
Cubic capacity/cylinder	791 cc/4 cylinder
hp/kW	70/51
Construction period	1966 – 1970
Number of produced cars	approx. 11.400

Honda S 800 Coupé

The Honda S 800 Coupé, an aggressive, noisy and speedy little car, was based on a backbone chassis with a wheel base of 200 centimetres. In the front as well as the rear, the model, which was only 320 centimetres long, had independent suspensions. Instead of the standard four-gear transmission with an asynchronous first gear, a five-speed gearbox was available at a surcharge.

Model	Honda S 800 Coupé
Cubic capacity/cylinder	791 cc/4 cylinder
hp/kW	70/51
Construction period	1966 – 1970
Number of produced cars	approx. 11.400

Mazda Cosmo 110

In the early 1960s, Tsuneji Matsuda, son of the legendary Mazda founder, initiated contact to Germany and visited the brilliant engineer Felix Wankel at the NSU plant in Neckarsulm. The ramifications of this visit continue to this day – the rotary engine and Mazda have belonged together since then. At Mazda, this technology was used for the first time in the Cosmo Sport Premiere.

Model	Mazda Cosmo 110
Cubic capacity/cylinder	2 x 491 cc
hp/kW	110/81
Construction period	1967 – 1972
Number of produced cars	approx. 1450

Nissan Datsun DC 3

In 1954, at the first Tokyo Motor Show, Nissan showed a sporty little car whose design resembled to a great degree the cars that could be found and bought on the British isle. According to factory specifications, the little vehicle was called a "sports two-seater". However, it was also made with other body versions.

Model	Nissan Datsun DC 3
Cubic capacity/cylinder	750 cc/4 cylinder
hp/kW	18/13.2
Construction period	1952 – 1957
Number of produced cars	–

Nissan Datsun Sports 1500

On the occasion of the eighth Tokyo Motor Show in 1961, Nissan showed an open two-seater based on the technical foundation of the Bluebird sedan model. In contrast to the Bluebird, the "Sports 1500" convertible – also known as "Fairlady" – received a modern four-speed transmission. The 150 km/h-fast car was only available on the Japanese market.

Model	Nissan Datsun Sports 1500
Cubic capacity/cylinder	1488 cc/4 cylinder
hp/kW	71/52
Construction period	1961 – 1965
Number of produced cars	–

Nissan Datsun Sports 1600

In the course of model development, the Sport 1500 was equipped with a 1.6-litre engine as of 1965. This engine, which originated from the Bluebird sedan, was well-suited to the convertible that was now called "Sports 1600" and which was able to easily reach 165 km/h. Thanks to another output increase (96 hp), the models made as of 1967 even achieved up to 175 km/h.

Model	Nissan Datsun Sports 1600
Cubic capacity/cylinder	1595 cc/4 cylinder
hp/kW	90/66
Construction period	1965 – 1970
Number of produced cars	–

Nissan Datsun Sports 2000

After the Nissan Datsun Sports 1600, which had been launched in 1965 already, was firmly established in the Asian market, Datsun introduced the Sports 2000 model two years later. This was a particularly agile alternative that reached a high speed of 205 km/h. This car, which was still called "Sport" or "Fairlady" was also exported to the USA where it had to be established as a competitive model to the British roadsters.

Model	Nissan Datsun Sports 2000
Cubic capacity/cylinder	1982 cc/4 cylinder
hp/kW	150/110
Construction period	1967 – 1970
Number of produced cars	–

Nissan Datsun 240 Z

German designer Albrecht Graf Goertz was foremost among the individuals who were responsible for providing the 240 Z with its attractive appearance. In 1963, he created the first prototype of the car. However, the project was quickly abandoned and was not taken up until the middle of the year 1966. While Goertz was no longer working for Nissan at that time, many of his ideas were implemented in retrospect.

Model	Nissan Datsun 240 Z
Cubic capacity/cylinder	2393 cc/6 cylinder
hp/kW	130/96
Construction period	1969 – 1974
Number of produced cars	–

Nissan Datsun 240 Z

At Nissan the letter Z signifies a sport car tradition spanning more than 35 years which by now consists of the fifth generation. The "big bang" of the subsequently world's most successful sports car series took place in 1969 with the debut of the Datsun 240 Z at the Tokyo Motor Show. The sleek two-seater was accompanied by the slogan "A coupé with comfort and boot space".

Model	Nissan Datsun 240 Z
Cubic capacity/cylinder	2393 cc/6 cylinder
hp/kW	130/96
Construction period	1969 – 1974
Number of produced cars	–

Nissan Datsun 240 Z

For export purposes, Nissan equipped the 240 Z with a 2.4-litre straight six-cylinder engine (140 hp). Thus the car was given a respectable output. Testers reported maximum speeds of approximately 200 km/h and acceleration to a speed of 100 km/h in less than ten seconds. The chassis of the 240 Z with independent suspension and radial tyres on steel rims corresponded exactly to the preferences of the time.

Model	Nissan Datsun 240 Z
Cubic capacity/cylinder	2393 cc/6 cylinder
hp/kW	140/103
Construction period	1969 – 1974
Number of produced cars	–

Nissan Datsun 240 Z 4

The strongest argument for the new sports car from the Far East was its sensationally affordable price of USD 3526. In comparison: the Volvo 1800 E cost USD 4500, a Corvette USD 5000 and a Jaguar E-Type even USD 5800. Practically from the first day of sales there were long waiting lists, and commercially apt retailers occasionally liked to add 500 dollars of "top surcharge" to its price.

Model	Nissan Datsun 240 Z 4
Cubic capacity/cylinder	2393 cc/6 cylinder
hp/kW	140/103
Construction period	1969 – 1974
Number of produced cars	–

Toyota Sports 800

Japanese designer Shozo Sato designed a little sports car which was introduced in 1965 as the Toyota Sports 800. The model, powered by an air-cooled boxer engine, was built exclusively for the Japanese market. However, the performance of the 800 did not correspond to its sleek exterior, as it took it a total of 13.3 seconds to accelerate from 0 to 100 km/h.

Model	Toyota Sports 800
Cubic capacity/cylinder	790 cc/2 cylinder
hp/kW	49/36
Construction period	1965 – 1979
Number of produced cars	approx. 3300

Toyota 2000 GT

As opposed to early Japanese automobile design, the Toyota 2000 GT displayed in 1967 a definitely Western European shape. The 2000 GT, which already gave an expression of speed while standing still, easily achieved a top speed of 215 km/h. It was powered by a six-cylinder-engine with two overhead camshafts that was, oddly enough, built by Yamaha!

Model	Toyota 2000 GT
Cubic capacity/cylinder	1988 cc/6 cylinder
hp/kW	150/110
Construction period	1967 – 1970
Number of produced cars	351

GT NM 4

1970
2000

Intelligent technology and modern design

En route to better performance

The longing for sporty elegance quickly got the upper hand in the 1970s, 1980s and 1990s – continuously and at regular intervals, all the cars that today are considered automotive history classics were presented at the international automotive fairs. More than ever before, there was demand for dignified two-door models, even if they were originally based on practical sedans. The most important thing above all else was that their design was created on the drawing board of a famous designer.

The fact that the looks had to be right, was also known to the Japanese. When Japanese cars entered the European markets for the first time in the early 1970s, they were initially not taken seriously. In the meantime however, they have established themselves in the sports car sector and recognised that they can only succeed with "Western" design styles. Incidentally, Saturo Nozaki, one of the leading Far Eastern automobile designers, was familiar with the European style of design from the start – after all, he was educated in the early 1960s at the "Art Center College of Design" in Los Angeles.

In the 1950s, a top speed of 180 km/h was an absolutely satisfactory value for a sports car. Twenty years later, it was expected to be a little bit higher – and the "output race" began. Even though numerous smaller cars by some manufacturers were extremely agile with a mere 1.3-litre four-cylinder engine and were actually top sellers, other manufactur-

ers insisted on a twelve-cylinder engine with four overhead camshafts for image reasons. The technology that was derived from racing resulted in a great amount of horsepower, enough for a compelling maximum speed of 250 km/h or more. At the same time, an end to this trend was not in sight. During the following years, the achievement curve continued to rise even more steeply.

As of the 1980s, the pressure to produce permanently higher output coupled with higher top speeds resulted in a growing number of unconventional body designs. Another unusual phenomenon was the increased use of so-called high tech materials. The new concepts and the unconventional appearance were frequently at odds, but were generally favourably accepted. The past few decades of automobile history have produced an-

other novelty-sports cars featuring four wheel drive technology. This technology, had already been developed several decades earlier, was applied in 1980 to the first Audi quattro. The winning concept, successfully applied in racing, had a lasting effect on the world of technology and was the basis of a success story that still continues today. The myth of the permanent propulsion of all four wheels continues to be stronger and more vivid than ever before. The technology has impressively established itself not only at Audi or in racing, but also on the street. Today, four-wheel drive signifies more than permanent traction – the concept has become the synonym for forceful vitality, for fast and confident driving.

For quite some time, the vehicle class of so-called sports limousines was also commercially very lucrative for automotive manufacturers. The BMW M5 can possibly be considered to be the trailblazer of this visually mostly unobtrusive species. In 1984, Lancia introduced the smooth saloon "Thema 8.32", which was likewise a high-calibre four-door model. Today, almost every manufacturer includes a four-door "high end version" with a noble interior in its portfolio, to the great delight of automobile connoisseurs.

Anyone who turns the ignition key or presses the start button of a modern sports car today can literally feel the strength of the engine. For some, this was a very emotional process – maybe even a quite personal "love affair" between car and driver.

Audi Quattro

In 1980, the first Audi Quattro stood in the limelight of the Geneva International Motor Show. It was the basis of a success story that continues to this day. The square-edged coupé immediately became a best-seller, because with its permanent four-wheel drive and the strong five-cylinder turbo engine, the car provides a sporty performance in a fascinating revolutionary way.

Model	Audi Quattro
Cubic capacity/cylinder	2144 cc/5 cylinder
hp/kW	200/147
Construction period	1980 – 1991
Number of produced cars	11.429

Audi Sport Quattro

In 1984, appeared within the Quattro series a special model built in a limited edition that enjoys a legendary reputation today – the Sport Quattro with a wheel base of only 2204 millimetres. Its newly developed four-valve turbo engine with aluminium-cylinder block generated a high speed of 306 hp, while the generous use of Kevlar and other lightweight construction materials turned it into a rally car for the street.

Model	Audi Sport Quattro
Cubic capacity/cylinder	2144 cc/5 cylinder
hp/kW	306/225
Construction period	1984
Number of produced cars	224

Audi Quattro

During the first years, in addition to an increasingly noble interior, only small technical changes were introduced to the "original Quattro". With the model development of 1987, the car received a slightly modified five-cylinder engine, which initiated its torque a little earlier. In 1989, the output rose through a new cylinder head to 220 hp, and the maximum speed now amounted to 230 km/h.

Model	Audi Quattro
Cubic capacity/cylinder	2144 cc/5 cylinder
hp/kW	200/147
Construction period	1980 – 1990
Number of produced cars	11.429

Audi Coupé S 2

As of 1990, the "S Models" with their Quattro engines considerably contributed to the further establishment of the dynamic image that Audi had acquired in racing. The forerunner of this model range was the Coupé S 2, the designated successor of the original Quattro. Its well-groomed character proved that sportiness and cultivation could harmonise ideally.

Model	Audi Coupé S 2
Cubic capacity/cylinder	2226 cc/5 cylinder
hp/kW	220/162
Construction period	1990 – 1995
Number of produced cars	–

Audi Quattro Spyder

Many Audi customers would have gladly acquired this car. However, the elegant two-seater Spyder was merely built for exhibition purposes. It was powered by a V6 mid engine placed horizontally behind the seats. This type of arrangement guarantees an optimum weight distribution. Thanks to its aluminium body, the Spyder topped the scales at a mere 1100 kilograms.

Model	Audi Quattro Spyder
Cubic capacity/cylinder	2771 cc/6 cylinder
hp/kW	174/128
Construction period	1991
Number of produced cars	2

Audi AVUS Quattro

An interesting "instrument" employed by Audi to maintain the fascination with the "Quattro", was the introduction of spectacular concept cars. In 1991, Audi presented in sequence at the fairs in Frankfurt and Tokyo two sports cars with four-wheel drive – the "Quattro Spyder" and the "AVUS Quattro," The latter featured a body made of polished aluminium sheet and accelerated to 100 km/h in a maximum of three seconds.

Model	Audi AVUS Quattro
Cubic capacity/cylinder	5998 cc/12 cylinder
hp/kW	509/374
Construction period	1991
Number of produced cars	unique piece

Audi TT Coupé

It took the Audi TT only about three years to go from an International Motor Show concept car in 1995 to the start of serial production in 1998. Compact, with concise lines and a clear design, it represents an Audi design in its most elementary form. Two 1.8-litre engines with five-valve technology and an exhaust turbocharger ensured that the sporty output matched the sporty look.

Model	Audi TT Coupé
Cubic capacity/cylinder	1781 cc/4 cylinder
hp/kW	180/132
Construction period	1998 – 2005
Number of produced cars	–

Audi TT Coupé Quattro

In 1999, the TT coupé and the TT roadster also became available as Quattros. With their transversely built-in engines and permanent four-wheel drive it takes them 6.4 seconds to go from 0 to 100 km/h and they can achieve top speeds of 243 km/h. Incidentally in 2004, 58 percent of all TT coupé cars and 42 percent of TT roadsters were delivered with four-wheel drives.

Model	Audi TT Coupé Quattro
Cubic capacity/cylinder	1781 cc/4 cylinder
hp/kW	225/165
Construction period	1999 – 2005
Number of produced cars	–

Audi TT Roadster

The first series of Audi TT came off the conveyor belt until the year 2005. It was based on the shortened platform of the VW Golf /Audi A3, one more reason why the two-seater with narrow dimensions was available at a realistic price. A big recall action in late 1999, which cost Audi EUR 100 million, did not diminish the popularity of the car.

Model	Audi TT Roadster
Cubic capacity/cylinder	1781 cc/4 cylinder
hp/kW	180/132
Construction period	1999 – 2005
Number of produced cars	–

Audi TT Roadster Quattro

Whether as a roadster or a coupé, the TT undoubtedly has given a new sharpness to the image of Audi. Within the scope of model development, the optical and technical retouching was rather limited, adding nothing that could have destroyed the car's flawless shape. The second TT generation debuted in the summer of 2006, starting with the launch of the coupé.

Model	Audi TT Roadster Quattro
Cubic capacity/cylinder	1781 cc/4 cylinder
hp/kW	225/165
Construction period	1999 – 2005
Number of produced cars	–

BMW 3.0 CSi

After the successful start of the coupé series with the BMW 2000 C in 1965, model development soon took its course. Because many customers preferred very powerful versions BMW in 1971 supplemented the offer with the 3.0 CSi. Under the bonnet of this coupé, a six-cylinder-injection engine was at work whose output provided the car with a maximum speed of 220 km/h.

Model	BMW 3.0 CSi
Cubic capacity/cylinder	2985 cc/6 cylinder
hp/kW	220/161
Construction period	1971 – 1975
Number of produced cars	–

BMW M 635 CSi

In the mid-1970s, BMW presented the 6 series – a supreme class coupé which soon sold exceptionally well. Until the end of production in 1989, more than 86,000 units (all types) were sold. This made this series more successful than any BMW coupé ever before. Incidentally, among the most popular model types was the relatively expensive M 635 CSi, which was hard to find even at that time.

Model	BMW M 635 CSi
Cubic capacity/cylinder	3453 cc/6 cylinder
hp/kW	286/210
Construction period	1984 – 1989
Number of produced cars	5915

BMW M 635 CSi

The BMW M 635 CSi – the strongest variation of the 6 series of the time – rolled off the conveyor belt from 1984 to 1989. The concept of this model was, basically, to equip the luxurious 2 + 2 seater coupé with the legendary M1 engine that had been slightly modified. The result of this synthesis was certainly worthwhile. Within a short period of time, the coupé with a high speed of 255 km/h (acceleration from 0 to 100 km/h in 6.4 seconds) found eager followers.

Model	BMW M 635 CSi
Cubic capacity/cylinder	3453 cc/6 cylinder
hp/kW	286/210
Construction period	1984 – 1989
Number of produced cars	5915

BMW 850 i

After BMW removed the 6-series coupé from the program in early 1989, they already had a successor ready for launch – the 850 i. The deluxe coupé with initially 300 hp was built until 1999 with many different output levels. At that time the first development model shown here was available with a starting price of DM 135,000. The maximum speed of the 850 i, by the way, was electronically reduced to 250 km/h.

Model	BMW 850 i
Cubic capacity/cylinder	4988 cc/12 cylinder
hp/kW	300/220
Construction period	1989 – 1992
Number of produced cars	18.513

BMW 850 CSi

With more than 31.000 sold units (all model types), the BMW 8 series – optically a succession model of the 6 series – constituted a special landmark in the BMW coupé history. Its eight- and twelve-cylinder engines moved the sleek car at speeds up to 250 km/h and allowed the driver to pleasurably glide along in a way that had never been made possible before.

Model	BMW 850 CSi
Cubic capacity/cylinder	5379 cc/12 cylinder
hp/kW	326/240
Construction period	1982 – 1996
Number of produced cars	1510

BMW Turbo

In 1972, the year of the Olympic games, the BMW Turbo was created, which was best described as a research laboratory on wheels. This design concept car was equipped with gull-wing doors and conceived as a pure two-seater. As a unique feature, it had a warning device for the braking distance as well as security distance to the car in front. The Turbo engine fitted into the plastic bodywork allowed a top speed of 250 km/h.

Model	BMW Turbo
Cubic capacity/cylinder	1990 cc/4 cylinder
hp/kW	280/206
Construction period	1972
Number of produced cars	?

BMW 2002 Turbo

There could hardly have been a more unfavourable time for the launch of this car than the international oil crisis. The BMW 2002 Turbo, an expensive car for enthusiasts debuted in 1973. The insider model with a top speed of 210 km/h (available starting at DM 18,720) remained in production for only ten months. Today, the car that was distinguished by its bug spoiler was an extremely sought-after classic.

Model	BMW 2002 Turbo
Cubic capacity/cylinder	1990 cc/4 cylinder
hp/kW	170/125
Construction period	1973 – 1974
Number of produced cars	1672

BMW M1

When the Paris Motor Show opened its gates in 1978, sports car fans had only one destination – the fair exhibit of the BMW Motorsport GmbH. This was where the BMW M1 was found, Germany's fastest street sports at that time. It was only 1140 millimetres high, was equipped with a 277 hp engine and considerably faster than 260 km/h. The elegant sports device was available, incidentally, for exactly DM 100.000!

Model	BMW M1
Cubic capacity/cylinder	3453 cc/6 cylinder
hp/kW	277/204
Construction period	1978 – 1981
Number of produced cars	450

BMW M 535 i

The BMW 5 series was launched following the end of the Olympic Games in 1972. Until 1981, the early 5 series generation was available in six output levels – while the highlight of the series was the sporty M 535 i with 218 hp. It was based on a reinforced chassis and equipped with a five speed sports gear in addition to internally ventilated disc brakes.

Model	BMW M 535 i
Cubic capacity/cylinder	3453 cc/6 cylinder
hp/kW	218/160
Construction period	1980 – 1981
Number of produced cars	1410

BMW M1

Although the M1 was a pure-bred sports car, its owners could enjoy a certain degree of comfort (small boot, air-conditioning, etc.). However, the M1 required quite some expertise to drive. As typical for mid-engine cars, in case of too much lateral acceleration, the rear could only be controlled by quick and strong counter steering.

Model	BMW M1
Cubic capacity/cylinder	3453 cc/6 cylinder
hp/kW	277/204
Construction period	1978 – 1981
Number of produced cars	450

BMW M5

As a successor of the M 535i, BMWs second sports saloon was ready in 1985 among retailers for test-drives. It was given the simple model name M5, the same abbreviation which was still in use today for the current construction pattern. Who wanted to acquire this wolf disguised in the modest sheep clothing of a four-door model, had to shell out a large amount of money – DM 80,750 were due for the M5 with the top speed of 245 km/h.

Model	BMW M5
Cubic capacity/cylinder	3453 cc/6 cylinder
hp/kW	286/210
Construction period	1985 – 1987
Number of produced cars	2180

BMW M5

Naturally, BMW also gave the third generation of M5 (as of 1987) an understated design. Its exterior was only distinguishable by a differently shaped spoiler, a large cooling air intake, light alloy wheels measuring 8 J x 18 in front and 9½ J x 18 in the rear, as well as four exhaust tail pipes. By the way, for this "Businessmen's Express" an optional car phone (C net) was available – at a cost of DM 8,300.

Model	BMW M5
Cubic capacity/cylinder	3535 cc/6 cylinder
hp/kW	315/232
Construction period	1988 – 1992
Number of produced cars	9892

BMW M5

Because the M5-models were from the start already derived for high driving dynamism, the engineers of the BMW Motorsport GmbH always had an excellent source on which to base the further development of this class. After the 340-hp M5 built between 1992 and 1996, they surprised in 1998 with a type with 400 hp and a V8 engine. Comment of a test driver: "A bundle of crude strength … ".

Model	BMW M5
Cubic capacity/cylinder	4941 cc/8 cylinder
hp/kW	400/294
Construction period	1998 – 2003
Number of produced cars	20 500

BMW M3

The M3 was presented for the first time to a wide general public at the Frankfurt International Motor Show (IAA) in autumn of 1985. Even without a special paintjob, it could be easily distinguished from the other models of the 3 series. Six inches above the boot lid, a spoiler featured prominently across the whole width of the car. Upon closer inspection, it also became apparent that the C column of the M3 was slightly wider and extended more flatly.

Model	BMW M3
Cubic capacity/cylinder	2302 cc/4 cylinder
hp/kW	200/147
Construction period	1986 – 1990
Number of produced cars	17.704

BMW M3

To find out how the M3 behaved in practice, everyone who had seen the car at the International Motor Show in 1985 had to be a little more patient. In spring 1986 the pilot series was ready and could be presented to the press in the appropriate setting at the racetrack of Mugello. The impressed testers found out that high calibre racing technology lay hidden underneath the beefy bodywork.

Model	BMW M3
Cubic capacity/cylinder	2302 cc/4 cylinder
hp/kW	200/147
Construction period	1986 – 1990
Number of produced cars	17.704

BMW M3 Evo

The stability of the M3 four-cylinder engine in the tough conditions of the racetracks resulted in a special offer for private customers in 1988. With the addition "Evo" (for "evolution") BMW introduced an especially powerful M3 with many spoiler details. A 220-hp engine powered this model, while the catalytic converter version even exhibited 215 hp.

Model	BMW M3 Evo
Cubic capacity/cylinder	2467 cc/4 cylinder
hp/kW	220/162
Construction period	1988 – 1991
Number of produced cars	500

BMW M3 Cabrio

In addition to the M3 Evo, which was produced in 1988 in a small series, BMW offered another surprise for convertible drivers: Finally, there was an open M3 based on the BMW 3 series. The 215-hp model reached a maximum speed of 239 km/h and was therefore among the fastest open four-seaters that were available for purchase in a small series.

Model	BMW M3 Cabrio
Cubic capacity/cylinder	2302 cc/4 cylinder
hp/kW	195/143
Construction period	1988 – 1991
Number of produced cars	786

BMW M3 Cabrio

Following the success of the BMW M3 convertible launched in 1988, BMW subsequently occupied this market niche with an even stronger variation three years later. Under its bonnet a six-cylinder engine was at work. The 250 km/h-fast car incidentally was serially equipped with an electromechanical convertible soft top system. For the "normal" 3 series convertible this practical extra was available only against a surcharge (DM 1600).

Model	BMW M3 Cabrio
Cubic capacity/cylinder	2990 cc/6 cylinder
hp/kW	286/210
Construction period	1991 – 1993
Number of produced cars	455

BMW M3 Sport Evolution

Until the end of 1992, a total of 17,970 M3 models of the first generation were produced by the factory of the BMW Motorsport GmbH, including almost 800 convertibles. Nobody had expected this success, either on the street or on the racetrack. The M3 has already been a popular collector's item for a long time, primarily the "Sport Evolution" version, built only 600 times, with its 238-hp engine.

Model	BMW M3 Sport Evolution
Cubic capacity/cylinder	2302 cc/4 cylinder
hp/kW	200/147
Construction period	1986 – 1990
Number of produced cars	17.704

BMW Z1

Ten years after the debut of the M1, the Z1 provided quite a sensation. The BMW Technik GmbH had originally conceived and built it as a technology bearer for alternative body designs. When the roadster finally went into production it was equipped with a steel monocoque chassis. While this construction was very light on the one hand, on the other hand it displayed great rigidity.

Model	BMW Z1
Cubic capacity/cylinder	2494 cc/6 cylinder
hp/kW	171/126
Construction period	1986 – 1991
Number of produced cars	8012

BMW Z1

With the BMW Z1, one could travel very quickly (225 km/h), because under its elegant bonnet sat a 2.5-litre inline six-cylinder engine. In addition, it featured a chassis that truly lived up to the sports car concept. Special driving pleasure however, could be derived from the fact that the doors (or rather the access hatches) could be completely lowered into the body.

Model	BMW Z1
Cubic capacity/cylinder	2494 cc/6 cylinder
hp/kW	171/126
Construction period	1986 – 1991
Number of produced cars	8012

BMW Z3 2.0i

The Z1 had hardly been removed from the program, when customers started to urgently call for a successor. BMW heeded this call in 1995 when the Z3, the first BMW originating in the USA debuted. It was produced exclusively at the factory in Spartanburg/South Carolina and was exported all over the world from there. In 1998, another special vehicle – the Z3 coupé – was added to the roadster.

Model	BMW Z3 2.0i
Cubic capacity/cylinder	1991 cc/6 cylinder
hp/kW	150/110
Construction period	1999 – 2000
Number of produced cars	–

BMW Z07

At the 1997 Tokyo Motor Show BMW presented a fascinating concept car, which indicated how the BMW 507, the sports car created in 1956, could have developed over the course of the time. The Z07 was based on a very modern technical design and therefore its spine consisted of an aluminium spaceframe. Its snubbers as well as parts of the rear were made of carbon fibre.

Model	BMW Z07
Cubic capacity/cylinder	4941 cc/8 cylinder
hp/kW	400/294
Construction period	1997
Number of produced cars	unique piece

BMW Z07

For historical tradition reasons, the only engine possible for the Z07 was a silky-soft running eight-cylinder engine which transmitted its power via a sequential six-gear transmission to the back wheels. The low built concept car with its elongated front part, incidentally, moved on Michelin-prototype tyres measuring 8 x 18 inches (front) and 9 x 18 inches (rear).

Model	BMW Z07
Cubic capacity/cylinder	4941 cc/8 cylinder
hp/kW	400/294
Construction period	1997
Number of produced cars	unique piece

BMW Z9 Gran Turismo

The BMW "Z9 Gran Turismo" was certainly among the main attractions of the 1999 Frankfurt International Motor Show (IAA). The message that BMW provided along with the Z9 was: "The world will become easier, not more complex for BMW drivers. Instead of a cornucopia of buttons, switchers and levers, the fascia of the Z9 Gran Turismo features only the start button and the light switch."

Model	BMW Z9 Gran Turismo
Cubic capacity/cylinder	3900 cc/8 cylinder
hp/kW	245/180
Construction period	1999
Number of produced cars	unique piece

BMW Z9 Gran Turismo

Under the ice-blue paint of the Z9 Gran Turismo there was a carbon shell and an aluminium space frame – similar to the Z8 BMW sports car. The coupé moves on wheels measuring 20 inches in front and 21 inches in the rear. Their size also indicates the power of 560 Newton meters inherent in the 3.9-litre V8 turbo-diesel engine with Common Rail direct injection.

Model	BMW Z9 Gran Turismo
Cubic capacity/cylinder	3900 cc/8 cylinder
hp/kW	245/180
Construction period	1999
Number of produced cars	unique piece

Ford Capri RS 2600

With a bonnet painted in matte black that discreetly highlighted the sporty nature of the RS version, the Capri I managed to conquer the hearts of the younger generation in the 1970s. While the least powerful model had to be content with a V4 engine (50 hp), the RS (six cylinders) offered real racing pleasure.

Model	Ford Capri RS 2600
Cubic capacity/cylinder	2550 cc/6 cylinder
hp/kW	150/110
Construction period	1970 – 1973
Number of produced cars	–

Ford Escort RS Cosworth

When the Ford Escort II series celebrated its debut in 1990, it initially consisted of the conventional average quality car for the average consumer. However, only two years later this model series was able to also please drivers with racing ambitions when the RS Cosworth became available! Incidentally, the car, with a top speed of 223 km/h was assembled by Karmann in Osnabrück.

Model	Ford Escort RS Cosworth
Cubic capacity/cylinder	1994 cc/4 cylinder
hp/kW	220/162
Construction period	1992 – 1996
Number of produced cars	–

Ford Escort XR3i

As part of the ongoing model development, the third Escort generation (1980 to 1986) was completely reworked and improved. In line with the previous models, once more a sporty version was released, the XR3i. While this car certainly enjoyed popularity among its fan community, it lacked the decisive final bite of the Gtis of the competition.

Model	Ford Escort XR3i
Cubic capacity/cylinder	1595 cc/4 cylinder
hp/kW	105/77
Construction period	1983 – 1985
Number of produced cars	–

Isdera Commendatore 112i

When the "Commendatore 112i" was launched in 1993, its price was around DM 400,000. For this amount, the buyer received a super sports car with a maximum speed of 342 km/h. In addition, under its bipartite engine hood in the rear it featured a Mercedes Benz engine. The body of the extremely low Commendatore (104 centimetres) rested on a stable tubular space frame and was made of glass fibre to save weight.

Model	Isdera Imperator 112i
Cubic capacity/cylinder	6900 cc/12 cylinder
hp/kW	620/403
Construction period	1993
Number of produced cars	–

Isdera Imperator 108i

The schwäbische Ingenieurgesellschaft für Styling Design und Racing mbH (Swabian Engineering Society for Styling, Design and Racing ltd. – Isdera) has been producing sports cars since 1983. All models are manually produced in tiny editions, a guarantee for the exclusivity of the brand. The Imperator 108i, whose gull-wing doors are reminiscent of the Mercedes-Benz-C 111 prototypes of the 1960s, celebrated its debut at the Geneva Car Salon of 1984.

Model	Isdera Imperator 108i
Cubic capacity/cylinder	4973 cc/8 cylinder
hp/kW	330/243
Construction period	1984–1990
Number of produced cars	–

Mercedes-Benz 350 SL

In 1968 already, Daimler-Benz decided that it would launch a new SL generation in 1971. Under the internal abbreviation R 107, the roadster was designed to the joy of its fans without an ugly roll bar, which would have destroyed its silhouette. Instead the only security measure in case of a potential roll-over was its sturdy A column.

Model	Mercedes-Benz 350 SL
Cubic capacity/cylinder	3499 cc/8 cylinder
hp/kW	200/147
Construction period	1971 – 1980
Number of produced cars	–

Mercedes-Benz 450 SL

A number of well-conceived details made driving the SL a pure pleasure. The readability of its three round instruments and the ergonomics of all operating elements were flawless. The seats were available from the start with headrests and automatic seat belts. Upon request, instead of the baggage space a kind of "jump seat" behind the seats could be ordered whose usefulness, however, was rather limited.

Model	Mercedes-Benz 450 SL
Cubic capacity/cylinder	4520 cc/8 cylinder
hp/kW	225/165
Construction period	1973 – 1979
Number of produced cars	–

Mercedes-Benz 280 SL

At the start of serial production, the new SL, for which a hard top-roof was optionally available, was resplendent with its 3.5-litre V8 engine that had already been developed in 1969. Its power helped the SL, weighing almost 1600 kilograms, to accelerate in a mere nine seconds from 0 to 100 km/h. Its maximum speed was 210 km/h – which meant it was 10 km/h faster than its predecessor model, the "Pagoden-SL".

Model	Mercedes-Benz 280 SL
Cubic capacity/cylinder	2746 cc/6 cylinder
hp/kW	185/136
Construction period	1974 – 1985
Number of produced cars	–

Mercedes-Benz 500 SL

In its not initially planned 18 years of service life, the SL (R 107) was fitted with a whole range of six- and eight-cylinder engines. The model names were accordingly varied. A total of exactly 237,287 SL cars (all versions included) came off the factory's conveyor belts. Incidentally, the 280 SL and 300 SL (as of 1985) belonged to the successful "6-cylinder faction".

Model	Mercedes-Benz 500 SL
Cubic capacity/cylinder	4973 cc/8 cylinder
hp/kW	240/176
Construction period	1980 – 1989
Number of produced cars	

Mercedes-Benz 300 SL

The eight-cylinder engine versions of the SL were initiated by the 350 SL (later 300 SL), followed by its bigger brother, the 450 SL. In 1980, appeared the 500 SL with a light alloy engine, which remained in the program from 1985, after reworking and increasing the output, until the discontinuation of this SL class in 1989. The 560 SL remained exotic in some way, since it was created only for the US market.

Model	Mercedes-Benz 300 SL
Cubic capacity/cylinder	2960 cc/6 cylinder
hp/kW	190/140
Construction period	1985 – 1989
Number of produced cars	–

Mercedes-Benz 350 SLC

In October 1971, Daimler-Benz surprisingly presented at the Paris Motor Show a coupé version based on the SL, the SLC. While until that time the high-class coupé cars had always been based on suitable sedans, a roadster was for the first time the basis for the implementation of the closed model version.

Model	Mercedes-Benz 350 SLC
Cubic capacity/cylinder	3499 cc/8 cylinder
hp/kW	200/147
Construction period	1971 – 1979
Number of produced cars	–

Mercedes-Benz 280 SLC

Although the basic shape of the SL roadster had been essentially maintained in the layout of the SLC coupé, the result was not always enthusiastically received. There were some details that simply did not suit the SLC. This was primarily due to the wheel base that had been extended by 36 centimetres – this was the only way that allowed comfortable seating also in the back row of the SLC.

Model	Mercedes-Benz 280 SLC
Cubic capacity/cylinder	2746 cc/6 cylinder
hp/kW	185/136
Construction period	1974 – 1981
Number of produced cars	–

Mercedes-Benz 380 SLC

To somewhat conceal the slightly unfortunate proportions of the SLC, the designers sketched a kind of lamella curtain for the rear part of the side windows. In the opinion of the specialist press however, this even further emphasised the lines of the long roof part. SLC owners seemed hardly disturbed by this fact, they primarily appreciated the technical qualities of their car.

Model	Mercedes-Benz 380 SLC
Cubic capacity/cylinder	3818 cc/8 cylinder
hp/kW	218/160
Construction period	1979 – 1981
Number of produced cars	–

Mercedes-Benz 450 SLC

In terms of technical improvements, the SL and SLC initially proceeded side by side. For example, as of 1973, the stronger V8 engine was available with 4.5 litres of cubic capacity. However, the output spiral soon turned the other way, at the start of the international oil crisis. Because customers wanted less fuel consuming engines, the majority of the buyers now favoured an in-line six-cylinder engine with 2.8 litres of cubic capacity (Type 280 SLC).

Model	Mercedes-Benz 450 SLC
Cubic capacity/cylinder	4520 cc/8 cylinder
hp/kW	225/165
Construction period	1972 – 1979
Number of produced cars	–

Mercedes-Benz 450 SLC 5.0

The fact that – in spite of high costs of petrol – thrift was not everything became apparent in 1977. That year, the coupé-top model became the absolute top output model. Under its bonnet it now contained a five-litre V8 engine with 240 hp. To ensure that the engine did not have to support a body that was too much of a burden, the hood, boot lid and bumpers were made of aluminium.

Model	Mercedes-Benz 450 SLC 5.0
Cubic capacity/cylinder	5025 cc/8 cylinder
hp/kW	240/176
Construction period	1977 – 1979
Number of produced cars	–

Mercedes-Benz 500 SLC

In the year 1980, SL and SLC received a new, very high-quality interior styled after the S-class saloon. Coupé buyers could benefit from this change for only a short time however, because the production of the SLC was ceased in 1981. Nevertheless, they could console themselves with the fact that they were allowed to drive the fastest and most powerful SLC of all times.

Model	Mercedes-Benz 500 SLC
Cubic capacity/cylinder	4973 cc/8 cylinder
hp/kW	240/176
Construction period	1979 – 1981
Number of produced cars	–

Mercedes-Benz 300 SL

In autumn 1989, Mercedes-Benz continued the tradition of SL roadsters by introducing a stylistically fascinating and technically demanding sports car. The series, which was first introduced at Geneva Motor Show – internally known as R 129 – represented the fourth SL generation and was produced from 1989 until July 2001 about 180,000 times (including all model variations).

Model	Mercedes-Benz 300 SL
Cubic capacity/cylinder	2960 cc/6 cylinder
hp/kW	190/140
Construction period	1989 – 1993
Number of produced cars	–

Mercedes-Benz SL 320

Thanks to a highly-developed mechanism, the automatic convertible top of the R 129 series opens and closes within time of a red light in traffic, allowing its occupants to enjoy every ray of sun. A wind deflector behind the seats, which was specifically developed for the SL, provided nearly draught-free open driving even at higher speeds.

Model	Mercedes-Benz 320 SL
Cubic capacity/cylinder	3199 cc/6 cylinder
hp/kW	231/170
Construction period	1993 – 1998
Number of produced cars	–

Mercedes-Benz 300 SL-24

The serial security package of the two-seater included a sensor-controlled roll bar that was automatically deployed within 0.3 seconds via spring tension and hydraulics in case of a crash or an extreme driving situation. The A column was reinforced on the inside by a tube and expediently complements the protective effect of the unique automatic roll bar.

Model	Mercedes-Benz 300 SL-24
Cubic capacity/cylinder	2960 cc/6 cylinder
hp/kW	231/170
Construction period	1989 – 1993
Number of produced cars	–

Mercedes-Benz 600 SL

To guarantee a dynamic sports car feel, the manufacturer equipped the SL with powerful six-, eight- and twelve-cylinder-engines. The output range of the engines extended from 245 hp to 500 hp. The engine program was headed by the quietly running twelve-cylinder version whose dual turbochargers and air-to-water intercooler made it one of the world's most modern passenger car engines.

Model	Mercedes-Benz 600 SL
Cubic capacity/cylinder	5987 cc/12 cylinder
hp/kW	394/290
Construction period	1992 – 1993
Number of produced cars	–

Mercedes-Benz SL 500

The standard equipment of the SL corresponded precisely to the wishes of demanding drivers who valued comfort. Automatic climate control, Xenon headlights, leather upholstered integral seats and much more were standard features of the whole model range. By the way, in 1993 the "SL" was placed before the model name, thus for instance the 500 SL became a SL 500.

Model	Mercedes-Benz SL 500
Cubic capacity/cylinder	4966 cc/8 cylinder
hp/kW	306/225
Construction period	1998 – 2001
Number of produced cars	–

Mercedes-Benz SL 600

The V12 engine of the SL 600 already developed its maximum torque of 800 Newton meters at 1800 rpm – this guaranteed sovereignty in every driving situation. For the acceleration from 0 to 100 km/h, the SL 600 only required 4.7 seconds. For the intermediate spurt from 60 to 120 km/h it required 4.9 seconds, while the maximum speed was electronically limited to 250 km/h.

Model	Mercedes-Benz SL 600
Cubic capacity/cylinder	5987 cc/12 cylinder
hp/kW	394/290
Construction period	1992 – 2001
Number of produced cars	–

Mercedes-Benz CLK-GTR

Based on the Mercedes CLK coupé, Daimler-Benz developed a top performance sports car that was at the same time the basis of a new racing vehicle for the FIA GT Championship. A limited street-legal version also emerged as a spin-off of this super coupé.

Model	Mercedes-Benz CLK-GTR
Cubic capacity/cylinder	6900 cc/12 cylinder
hp/kW	560/412
Construction period	1997 – 2000
Number of produced cars	–

Mercedes-Benz SLK 200

With the legendary 190 SL, Mercedes-Benz already proved 50 years ago that a compact roadster could offer great driving pleasure. In 1996, the Stuttgart automobile manufacturer once more followed up on the theme and presented the SLK. Right from the start, the SLK set new standards for this vehicle class: aesthetics, coupled with high-class technology and exemplary security.

Model	Mercedes-Benz SLK 200
Cubic capacity/cylinder	1998 cc/4 cylinder
hp/kW	136/100
Construction period	1996 – 2000
Number of produced cars	–

Mercedes-Benz CLK-GTR

The high-tech engine of the CLK-GTR offered a maximum torque of 720 Newton meters, which provided substantial acceleration and drive output. A racing experienced six-gear gearbox with sequential gear changing provided the power transmission. The engine was screwed rigidly to the back wall of the monocoque, and together with the gear system also provided a supporting function.

Model	Mercedes-Benz CLK-GTR
Cubic capacity/cylinder	6900 cc/12 cylinder
hp/kW	560/412
Construction period	1997 – 2000
Number of produced cars	–

Mercedes-Benz Vision SLR Coupé

The silver arrow of tomorrow – this was embodied by the SLR concept car that Mercedes-Benz introduced to the general public in 1999 at the Auto-Show in Detroit. The long stretched bonnet, the powerfully curved fenders and the gull-wing doors of the Vision SLR were based on stylistic concepts of the legendary SL models from the 1950s and their racing sport variations SLR with which the brand scored one victory after another at the time.

Model	Mercedes-Benz Vision SLR Coupé
Cubic capacity/cylinder	5500 cc/8 cylinder
hp/kW	557/410
Construction period	1999
Number of produced cars	unique piece

Mercedes-Benz Vision SLR Roadster

At the Frankfurt International Motor Show (IAA) of 1999, Mercedes-Benz presented the roadster variation of the Vision SLR concept car. It was closely based on the closed variation, yet set other accents, including for instance a hood made of a new light-permeable material. The road performance of the sports car consisted of the following – acceleration from 0 to 100 km/h in 4.2 seconds, top speed 320 km/h.

Model	Mercedes-Benz Vision SLR Roadster
Cubic capacity/cylinder	5500 cc/8 cylinder
hp/kW	557/410
Construction period	1999
Number of produced cars	unique piece

Opel Manta A GT/E

In 1970, the Manta A was launched, one of the most successful trend-setting cars ever. One reason for its big popularity was the smartly styled body – it made the Manta appear sporty. To be able to better position the car that was built at various output levels, the model series was described by the slightly awkward concept "family coupé".

Model	Opel Manta A GT/E
Cubic capacity/cylinder	1897 cc/4 cylinder
hp/kW	105/77
Construction period	1974 – 1975
Number of produced cars	–

Opel Manta A GT/E Black Magic

The first Manta series came off the conveyor belt from 1970 to 1975. The more powerful models SR and GT/E were in great demand. Its black bonnet and painted on rally stripes gave a very sporty impression. For the year 1975, Opel introduced the variation "Black Magic", a special model with particularly extensive equipment.

Model	Opel Manta A GT/E
Cubic capacity/cylinder	1897 cc/4 cylinder
hp/kW	105/77
Construction period	1975
Number of produced cars	–

Opel Manta SR

Slightly enlarged and optically enhanced, the second Manta generation was presented by Opel in the summer of 1975. The new edition corresponded exactly to the preferences of the time by featuring a deeper waistline and rectangular headlights. The well-gradated engine range was initially taken over almost without any changes from the Manta A.

Model	Opel Manta SR
Cubic capacity/cylinder	1897 cc/4 cylinder
hp/kW	90/66
Construction period	1975 – 1988
Number of produced cars	–

Opel Manta GT/E

"Dream of what makes you happy," was the promise of advertising campaign published on the debut of the Manta GT/E. Similar to its predecessor, the Manta A, an especially powerful engine was positioned under the bonnet of this model. Optional accessories included, for instance, wide tyres and light alloy rims, which gave the GT/E an even sportier look.

Model	Opel Manta GT/E
Cubic capacity/cylinder	1897 cc/4 cylinder
hp/kW	105/77
Construction period	1977 – 1988
Number of produced cars	–

Opel Ascona B 400

Specifically for application in-house, Opel transformed the actually quite plain Ascona B into a hot competitive model, with which Walter Röhrl won the rally world championship of that time. Private drivers were given a more civilised variation: Opel built for them the Ascona 400 with a maximum speed of more than 200 km/h with a throttled output.

Model	Opel Ascona B 400
Cubic capacity/cylinder	2400 cc/4 cylinder
hp/kW	140/103
Construction period	1979 – 1981
Number of produced cars	–

Opel Manta B 400

Opel presented in 1981 the Manta 400 as a successor to the Ascona 400. It appealed especially to drivers with racing ambitions and was equipped with a 16-valve engine with two overhead camshafts. A sports chassis, internally ventilated disc brakes as well as aluminium wheels with tyres measuring 205/60 VR 15 additionally highlighted the sporty character of the car.

Model	Opel Manta B 400
Cubic capacity/cylinder	2400 cc/4 cylinder
hp/kW	144/106
Construction period	1981 – 1984
Number of produced cars	–

Opel Kadett GSi

Basically, the Opel Kadett E was quite a normal, practical everyday use vehicle. It had a good price-performance ratio right from the start because its entrance-level engine consisted of a 55-hp four-cylinder variation. However, if the car was implanted with a 150-hp engine featuring a cylinder head optimised by Lotus, the front-wheel drive vehicle easily reached 200 km/h.

Model	Opel Kadett GSi
Cubic capacity/cylinder	1998 cc/4 cylinder
hp/kW	150/110
Construction period	1988 – 1991
Number of produced cars	–

Opel Lotus Carlton

When Lotus still belonged to the General Motors Group, the extremely tuned Carlton was created on the basis of the Opel Omega 3.0 GSi, a certified wolf in sheep's clothing. The car, which was painted exclusively in "Imperial-Green", was sold on the British market about 320 times by Vauxhall, while Opel could sell 630 Carlton models. The original plans included the sale of 1100 units.

Model	Opel Lotus Carlton
Cubic capacity/cylinder	3595 cc/6 cylinder
hp/kW	377/277
Construction period	1989 – 1992
Number of produced cars	950

Opel Calibra 2.5i

Attractive styling combined with a favourable cw-value distinguished the Opel Calibra. The 2+2-seater coupé built for seven years was available in several output levels (115 hp to 204 hp) including an all-wheel-drive version. The series that was initially equipped with a four-cylinder engine profited later from the V6 engine of the Opel Omega.

Model	Opel Calibra 2.5i
Cubic capacity/cylinder	2498 cc/6 cylinder
hp/kW	170/125
Construction period	1993 – 1997
Number of produced cars	–

Porsche Carrera RS 2.7

In 1972, the Carrera logotype was featured for the first time on a Porsche 911. The name came from the Carrera Panamericana, a street race through Mexico that was held in the 1950s. The 911 Carrera RS 2.7 with its 210-hp engine quickly became Germany's fastest street – legal car. Its maximum speed was 245 km/h. A distinguishing mark of this model was its "duck's rump" placed on the rear.

Model	Porsche Carrera RS 2.7
Cubic capacity/cylinder	2687 cc/6 cylinder
hp/kW	210/154
Construction period	1973 – 1975
Number of produced cars	

Porsche 911 E

The most important modifications of Porsche 911 in the course of the model years until 1973 included for instance, the introduction of a Targa model. In addition, a more affordable version called Porsche 912 enriched the market. In terms of engine technology, the variations 911 T, E and S were distinguished. Another notable characteristic of this decade was the cubic capacity expansion from 2.0 to 2.2 litres and, finally, 2.4 litres.

Model	Porsche 911 E
Cubic capacity/cylinder	2341 cc/6 cylinder
hp/kW	165/121
Construction period	1970 – 1973
Number of produced cars	–

Porsche 911 Turbo 3.0

Automotive historians call the years between 1973 and 1989 the "G series" years in the history of the Porsche 911. From initially 2.7 litres, the cubic capacity increased to 3.2 litres, for the Turbo even to 3.3 litres. In addition, the name Carrera appeared in connection with 911. This was for the first time in 1972 for the especially sporty model Carrera RS 2.7, a basic vehicle for racing applications.

Model	Porsche 911 Turbo 3.0
Cubic capacity/cylinder	2994 cc/6 cylinder
hp/kW	260/191
Construction period	1974 – 1976
Number of produced cars	–

Porsche 911 Turbo 3.3

In 1974, Porsche introduced the world's first standard sports car featuring exhaust turbo-supercharging. The enterprise also introduced another unusual feature – Porsche was the first manufacturer to serially employ dual-sided hot-dip galvanised steel metal for the bodywork. Initially Porsche granted a six-year guarantee against rust penetration that was raised to ten years in 1985.

Model	Porsche 911 Turbo 3.3
Cubic capacity/cylinder	3299 cc/6 cylinder
hp/kW	300/221
Construction period	1976 – 1990
Number of produced cars	–

Porsche 911 SC 3.0 Cabrio

October 1982 was quite a special date for Porsche enthusiasts: Finally, they were once again presented with a fully convertible car. While there were some Targa models introduced in the meantime, however the construction of "real" convertibles had been ceased in 1965 with the Porsche 356 C 1600. In terms of engine technology, all SC versions were equipped, incidentally, with a modern K-Jetronik-injection system.

Model	Porsche 911 SC 3.0 convertibles
Cubic capacity/cylinder	2994 cc/6 cylinder
hp/kW	204/150
Construction period	1982 – 1983
Number of produced cars	–

Porsche 911 Turbo 3.3

The 911 Turbo 3.0 was the first standard vehicle whose increase in output was achieved with the help of an exhaust turbo supercharger. In 1977, the Porsche turbo technology was complemented with an intercooler, which raised the density of the air while keeping the pressure constant. With this new technology and an increase of the cubic capacity to 3.3 litres, the output of the Porsche, now called 911 Turbo 3.3 increased to 300 hp at 5500 rpm.

Model	Porsche 911 Turbo 3.3
Cubic capacity/cylinder	3299 cc/6 cylinder
hp/kW	300/221
Construction period	1976 – 1990
Number of produced cars	–

Porsche 911 Carrera 3.2

Superior road performance distinguished the sports car from Zuffenhausen from the start. In 1963, this meant that it accelerated in 9.1 seconds from 0 to 100 km/h and achieved a maximum speed of 210 km/h. At that time as well as today, the engine consisted of a six-cylinder boxer's engine with an unmistakable sound, even though the output has in the meantime increased, depending on the model, to more than 420 hp.

Model	Porsche 911 Carrera 3.2
Cubic capacity/cylinder	3164 cc/6 cylinder
hp/kW	231/170
Construction period	1983 – 1989
Number of produced cars	–

Porsche 911 Carrera 2 Speedster

The Speedster was exclusively built by Porsche, following a tradition that began in 1954. The model was always built in small volumes, a fact that will not change in the future. After the success of the 1988 speedster, Porsche presented three years later a successor at the Paris Motor Show. The Speedster had a maximum speed of 260 km/h and accelerated to 100 km/h in 5.7 seconds.

Model	Porsche 911 Carrera 2 Speedster
Cubic capacity/cylinder	3164 cc/6 cylinder
hp/kW	231/170
Construction period	1988 – 1989
Number of produced cars	2103

Porsche Carrera 2 3.6 Targa

An output of 130 hp was quite respectable for a 2-litre engine in 1963. Nevertheless, the strength of the Porsche engines increased even further. Ferdinand Porsche said: "If I had known in 1960 that one could increase the engine to three litres and more, I would have said that it was needlessly big and heavy and would have instructed the engineers to make it smaller. I am glad that I did not know this fact!"

Model	Porsche Carrera 2 3.6 Targa
Cubic capacity/cylinder	3600 cc/6 cylinder
hp/kW	250/184
Construction period	1989 – 1993
Number of produced cars	–

Porsche 911 GT 2

In 1995, Porsche presented a new edition based on the "993" turbo model. The 3.6-litre engine that was now equipped with two turbochargers delivered an output of 408 hp. In the same year the first GT 2 also debuted. It consisted of a lightweight construction version of the Carrera model for street and racing applications with 430 to 600 turbo hp. For drivers with particularly sporty ambitions, the Carrera RS continued to be available.

Model	Porsche 911 GT 2
Cubic capacity/cylinder	3600 cc/6 cylinder
hp/kW	462/340
Construction period	1995 – 1996
Number of produced cars	–

Porsche 911 Targa 3.6

In 1995, a new Targa was also among the novelties of Porsche. The elegant model was technically based on the 911 Carrera, yet as opposed to earlier Targa versions the novice featured a wide glass roof. The two separately removable roof halves that were a feature of earlier models were no longer featured in this Targa.

Model	Porsche 911 Targa 3.6
Cubic capacity/cylinder	3600 cc/6 cylinder
hp/kW	272/200
Construction period	1995 – 1997
Number of produced cars	–

Porsche 911 GT 1

After the Porsche 911 GT 1 had achieved several racing victories in 1996, the manufacturer decided to also build this super sports car in an exclusive small series as a street version. Weighing only 1150 kilograms the GT 1 accelerated from 0 to 100 km/h in only 3.7 seconds and had a top speed of 310 km/h. The sports car that cost DM 1.5 million at that time was painted according to the wishes of the customers.

Model	Porsche 911 GT 1
Cubic capacity/cylinder	3163 cc/6 cylinder
hp/kW	544/400
Construction period	1997
Number of produced cars	30

Porsche 924

Porsche was well-advised to buy back the model 924, originally developed for VW, to complement its own model range. Soon it became apparent that the 924 resulted in remarkably high sales figures. While the sports car legend 911 required about twelve years to reach a production volume of 100.000 units, the smaller Type 924 reached this number in a little over five years.

Model	Porsche 924
Cubic capacity/cylinder	1984 cc/4 cylinder
hp/kW	125/92
Construction period	1976 – 1985
Number of produced cars	150.000

Porsche 924 S

Similar to the Porsche 928 that was developed at the same time, the 924 was also constructed with the transaxle method with the engine positioned in the front and the gear placed on the rear axle. The water-cooled inline engine used for the first time by Porsche (featuring an overhead camshaft and K-Jetronic-injection) accelerated exceptionally well and was considered rather moderate in fuel consumption (10 to 12 litres /100 kms).

Model	Porsche 924 S
Cubic capacity/cylinder	2479 cc/4 cylinder
hp/kW	150/110
Construction period	1985 – 1988
Number of produced cars	16.270

Porsche 924 Turbo

In November 1978, Porsche introduced the model 924 Turbo, a sensational development in terms of engine power and road performance. Its charged engine with an unchanged cubic capacity of two litres, had an output of 170 hp (125 kW). The car required only 7.0 seconds to accelerate from 0 to 100 km/h, while the maximum speed of the agile sports car was, fortunately, even a little higher than the stated 225 km/h.

Model	Porsche 924 Turbo
Cubic capacity/cylinder	1984 cc/4 cylinder
hp/kW	170/125
Construction period	1978 – 1985
Number of produced cars	12,420

Porsche 924 Carrera GTS

For customers with very sporty ambitions, Porsche offered as of 1980 the 924 Carrera GTS with intercooling. This 210-hp sports vehicle – recognisable by its expanded fenders and an air intake duct on the bonnet – received one year later a competitor from within the same make, the even more aggressive 924 Carrera GTS (245 hp). The tyres, chassis and brakes were especially adapted to the abilities of this car.

Model	Porsche 924 Carrera GTS
Cubic capacity/cylinder	1984 cc/4 cylinder
hp/kW	245/180
Construction period	1981
Number of produced cars	59

Porsche 928

When Porsche presented the 928 at the Geneva Motor Show in 1977, only insiders knew that the development work for this car had already begun in 1971. Because Porsche wanted to create a progressive vehicle, they decided on a modern construction method and made the doors and fenders of aluminium, while the bodywork was made according to tradition from galvanised steel sheets.

Model	Porsche 928
Cubic capacity/cylinder	4474 cc/V8 cylinder
hp/kW	240/176
Construction period	1977 – 1982
Number of produced cars	17.660

Porsche 928 S

In 1980, the Porsche 928 S, a type developed as part of model development became available for test drives at Porsche retailers. The enlargement of the engine drilling from 95 to 97 millimetres increased the cubic capacity to 4.7 litres. The eight-cylinder engine now had an output of 300 hp, and it took only 6.6 seconds for the speedometer needle to reach the 100 km/h mark, while the car's maximum speed was approximately 250 km/h.

Model	Porsche 928 S
Cubic capacity/cylinder	4664 cc/V8 cylinder
hp/kW	300/221
Construction period	1980 – 1986
Number of produced cars	22.660

Porsche 928 GT

The Porsche 928 was certainly not an economic car at an affordable price. Nevertheless, it was chosen by an internationally selected professional jury as the "Car of the year 1978." This was the first time that this coveted title could be claimed by a sports car. Incidentally, to reduce the certainly high fuel consumption of the car, the engine was give a higher condensation as of 1979.

Model	Porsche 928 GT
Cubic capacity/cylinder	4957 cc/V8 cylinder
hp/kW	330/243
Construction period	1989 – 1991
Number of produced cars	–

Porsche 928 S4

As of 1987, the 928 appeared with a more rounded look. This facelift in combination with a new lining of the underbody tremendously improved the aerodynamics of the car. Also in terms of engine technology, the 928 S4 presented itself as more agile than ever before, because the powerful eight-cylinder engine received new pistons and a revised cylinder head with optimised combustion chambers, among other improvements.

Model	Porsche 928 S4
Cubic capacity/cylinder	4957 cc/V8 cylinder
hp/kW	320/235
Construction period	1987 – 1991
Number of produced cars	

Porsche 928 GTS

As of 1992, the Porsche 928 GTS, successor of the models 928 S4 and 928 GT was ready for launch. With its body that was expanded in the rear area and the rear spoiler painted in the same colour as the body, the 275 km/h fast sports car matched the preferences of the era. The 928 GTS could be ordered alternatively with a wide ratio manual transmission or a close ratio automatic transmission.

Model	Porsche 928 GTS
Cubic capacity/cylinder	5397 cc/V8 cylinder
hp/kW	350/257
Construction period	1992 – 1995
Number of produced cars	–

Porsche 944

In constructing the light alloy V8 engine for the big Porsche 928, the engineers of the Porsche development centre in Weissach had an ulterior motive – from the start they pursued the plan of potentially halving the V8 into a straight four-cylinder engine and use it to drive the advancement of the 924. This was the clever way in which the engine of the Porsche 944, introduced in 1981, was derived.

Model	Porsche 944
Cubic capacity/cylinder	2479 cc/4 cylinder
hp/kW	163/120
Construction period	1981 – 1991
Number of produced cars	–

Porsche 944 Turbo S

Porsche 944 was among the cars that sold excellently right from the start. The customers accepted the price, which was already rather high for the basic model and also ordered diligently the model types for which even higher prices were charged. These included in addition to the 944 Turbo (starting at 220 hp, maximum speed 260 km/h) also the especially vicious 944 Turbo S of the model year 1988.

Model	Porsche 944 Turbo S
Cubic capacity/cylinder	2479 cc/4 cylinder
hp/kW	250/184
Construction period	1988
Number of produced cars	–

Porsche 944 S2

In 1990, the cubic capacity of the model 944 S2 increased to a considerable three litres (now the engine was fitted with four valves per cylinder). This meant for the coupé as well as for the convertible an output increase to 211 hp (155 kW) and a maximum speed of 240 km/h. The alternative model, the 944 Turbo, was powered by a 250-hp (184 kW) strong engine which allowed the cars to reach speeds of 260 km/h.

Model	Porsche 944 S2
Cubic capacity/cylinder	2990 cc/4 cylinder
hp/kW	211/155
Construction period	1990 – 1991
Number of produced cars	–

Porsche 944 S2 Convertible

It was said that the double balance shafts of the 944th engine lend this engine the smoothness and the character of a six-cylinder engine. This character was to kept intact in the course of technical developments. In 1985 debuted the model 944 Turbo with a 2.5 litre engine and initially 220 hp (162 kW). At that time its road performance was nearly equal to that of the Porsche 911.

Model	Porsche 944 S2 Convertible
Cubic capacity/cylinder	2990 cc/4 cylinder
hp/kW	211/155
Construction period	1989 – 1991
Number of produced cars	7000

Porsche 944 Turbo Cabrio

In addition to the coupé, Porsche built the 944 until 1991 also as an open model type. In terms of available space, both models were identical, however, on account of the hood construction the convertible provided a smaller trunk. As a highlight, in 1991 Porsche also produced the turbo-version in the convertible style. With a top speed of 260 km/h, this made the 944 one of the world's fastest convertibles of that time.

Model	Porsche 944 Turbo Cabrio
Cubic capacity/cylinder	2479 cc/4 cylinder
hp/kW	250/184
Construction period	1991
Number of produced cars	528

Porsche 959

The power output of the Porsche 959 (450 hp) model was sufficient to easily propel the race car to a maximum speed of 315 km/h. Those who wanted to call this model their own, had to pay quite an amount of money – the 959 cost DM 420,000. With the 959 Porsche wanted to present, incidentally, its visualisation of the sports car of the future.

Model	Porsche 959
Cubic capacity/cylinder	2849 cc/6 cylinder
hp/kW	450/331
Construction period	1987 – 1988
Number of produced cars	292

Porsche 959

The year 1987 was very important in the history of the Porsche company: it was the year in which quite a special model of the 911 series was introduced, the 959. The 959 was a "Super 911" and an all-wheel technological innovation. Initially intended for racing in the so-called "Group B", this high-capacity sports car was equipped with everything that was technically possible without economic restrictions.

Model	Porsche 959
Cubic capacity/cylinder	2849 cc/6 cylinder
hp/kW	450/331
Construction period	1987–1988
Number of produced cars	292

Porsche 968

Permanent advancements of the engine made the Porsche 968 a record holder in various aspects. For example, at that time, the 3-litre engine with its 240 hp (177 kW) output was considered to be the engine with the highest torque of its cubic capacity class. Unfortunately, the low production capacity in Zuffenhausen ended the career of 968 prematurely – it had to make room for a new successful model, the Porsche Boxster.

Model	Porsche 968
Cubic capacity/cylinder	2990 cc/4 cylinder
hp/kW	240/176
Construction period	1992 – 1994
Number of produced cars	

Porsche 968 Cabrio

The Porsche 968 introduced in 1991 was destined to become at the same time the climax and final point of the development of four-cylinder sports cars by Porsche. In the meantime, there was world of difference between the 968 and the original 924 – from the technology to the body design. While the types 924 and 944 had been built at the Audi factory in Neckarsulm, the 968 was constructed at the Porsche headquarters factory in Stuttgart-Zuffenhausen.

Model	Porsche 968 Cabrio
Cubic capacity/cylinder	2990 cc/4 cylinder
hp/kW	240/176
Construction period	1992 – 1994
Number of produced cars	–

Porsche Boxster

The Boxster, which combines the dynamic driving qualities of a sports car with unlimited suitability for everyday use, was introduced by Porsche in September 1996. It set new output and security standards in the market segment of open sports cars. New kinematics for the convertible soft-top allow the opening and closing of the electrically controlled top in the unparalleled time of only twelve seconds.

Model	Porsche Boxster
Cubic capacity/cylinder	2480 cc/6 cylinder
hp/kW	204/150
Construction period	1996 – 1999
Number of produced cars	–

Porsche Boxster S

The Boxster was distinguished from its competitors by its mid-engine concept of a six-cylinder Boxer engine typical of Porsche (2.5 litres of cubic capacity, 204 hp/150 kW) which requires in a practically-oriented driving mode less than ten litres of Super plus fuel for 100 kilometres. With a c_d value of 0.31 and a top speed of 240 km/h (from 0 to 100 km/h in 6.9 seconds) the Boxster was best of its class.

Model	Porsche Boxster S
Cubic capacity/cylinder	3179 cc/6 cylinder
hp/kW	252/185
Construction period	1999 – 2002
Number of produced cars	–

VW Karmann-Ghia TC

Unfortunately, the fast-looking Karmann Ghia TC was not a pure-bred sports car. It rather belonged to the category of deceptive packages, which were covered with a sporty-looking body. The TC, internally also known as "Minas," was produced exclusively for the South American market and was built by Karmann Ghia do Brasil.

Model	VW Karmann-Ghia TC
Cubic capacity/cylinder	1584 cc/4 cylinder
hp/kW	54/40
Construction period	1970 – 1975
Number of produced cars	18 000

Volkswagen Golf GTI Pirelli

Towards the end of the first Golf generation, Volkswagen launched the "Pirelli GTI" as a sort of thank you. The limited special model received extensive standard equipment including tinted windows, a 112 hp engine with a five-speed gearbox, leather sports steering wheel and much more. This GTI accelerated from 0 to 100 km/h in 9.2 seconds.

Model	VW Golf GTI Pirelli
Cubic capacity/cylinder	1781 cc/4 cylinder
hp/kW	112/82
Construction period	1983
Number of produced cars	10.500

Volkswagen Golf GTI

In 1976, a 182 km/h-fast and compact Volkswagen caused quite a stir when it appeared for the first time in the rear-view mirrors of fast sports cars and heavy sedans – the GTI. Planned production of this model was a limited volume of 5000 units. However, the GTI rose immediately to a best-seller and became synonymous of the sporty compact class of cars.

Model	VW Golf GTI
Cubic capacity/cylinder	1781 cc/4 cylinder
hp/kW	110/81
Construction period	1976 – 1983
Number of produced cars	350.000

Volkswagen Golf GTI

In the second generation introduced in 1984 the GTI continued to be a phenomenon within the VW brand. Although it had to share the market now with com petitors called GSi or XRi, the majority of the buyers opted for the "original" among the fast compact cars. The average annual production of the GTI was 630.000 units.

Model	VW Golf GTI
Cubic capacity/cylinder	1781 cc/4 cylinder
hp/kW	129/95
Construction period	1984 – 1990
Number of produced cars	–

Volkswagen Golf GTI G 60

In 1991, debuted the third GTI generation that was built until 1997. Statistically speaking, about 530.000 drivers opted for this car annually. Among the highlights of the third series was the GS 60 that was only produced for a short time and achieved a output of 160 hp with the help of an intercooler. It had a maximum speed of 220 km/h.

Model	VW Golf GTI G 60
Cubic capacity/cylinder	1781 cc/4 cylinder
hp/kW	160/118
Construction period	1991
Number of produced cars	–

Volkswagen Scirocco GTI

In 1974, even before the debut of the Golf, VW already launched its coupé version called Scirocco. After the new 2 + 2-seater had established itself on the market, the model series was complemented in 1976 with a GTI variation with a top speed of approximately 190 km/h. In late 1980, after the construction of about 495,000 built units (all models) the career of the first Scirocco generation ended.

Model	Volkswagen Scirocco GTI
Cubic capacity/cylinder	1781 cc/4 cylinder
hp/kW	110/81
Construction period	1976 – 1980
Number of produced cars	–

Volkswagen Scirocco 1.8 16V

Substantially more smoothly styled than its predecessor, the Scirocco II appeared on the streets starting in 1981. This model, designed by Giugiaro was equipped in the end with a 1.8-litre injection engine, either with two or with four valves per cylinder. The latter version accelerated the Scirocco after 8.6 seconds from 0 to 100 km/h, while its maximum speed was 200 km/h.

Model	Scirocco 1.8 16V
Cubic capacity/cylinder	1781 cc/4 cylinder
hp/kW	129/95
Construction period	1985 – 1992
Number of produced cars	–

Volkswagen Corrado 1.8 G 60

In terms of technology, the VW Corrado (produced 1988 until 1995) was based on the Golf II. The manufacturer envisioned the modern 2 + 2-seater with the big boot lid to become the successor of Scirocco. However, unfortunately, it belonged to a much higher prize category. The car with a maximum speed of 235 km/h was built with three output levels: 136 hp, 160 hp and 190 hp.

Model	Corrado 1.8 G 60
Cubic capacity/cylinder	1781 cc/4 cylinder
hp/kW	160/118
Construction period	1988 – 1993
Number of produced cars	–

Volkswagen W 12

At the Tokyo Motor Show in 1997, Volkswagen presented for the first time the "More than six-cylinder", Volkswagen W 12. The unusual shape of the W 12 stemmed from collaboration with Italdesign. Equally modern was the car's unusually compact twelve-cylinder engine. It was made in the space-saving W construction method where two VR6 engines are offset from each other at an angle of 72 degrees.

Model	Volkswagen W 12
Cubic capacity/cylinder	5584 cc/12 cylinder
hp/kW	420/309
Construction period	1997
Number of produced cars	unique piece

Volkswagen W 12

With the debut of the W 12, it was already decided that Volkswagen would continue to develop this unusual design over the course of time, because this was the only way to utilise the potential offered by this car. The engine of the version developed in 1997 was especially compact with a length of 51 centimetres and a width of 70 centimetres. Its crankcase with plasma-coated cylinder liner was made of aluminium.

Model	Volkswagen W 12
Cubic capacity/cylinder	5584 cc/12 cylinder
hp/kW	420/309
Construction period	1997
Number of produced cars	unique piece

Volkswagen W 12 Roadster

Following the presentation of the closed W12 concept car, one year later VW reintroduced the concept car again as an avant-garde roadster. The basic features of the engine system remained unchanged – a strong torque was achieved by an engine control with infinitely variable, hydraulic camshafts adjustment – making 520 Newton meters available at 3000 tours/minute.

Model	Volkswagen W 12 Roadster
Cubic capacity/cylinder	5584 cc/12 cylinder
hp/kW	420/309
Construction period	1997
Number of produced cars	unique piece

Volkswagen W 12 Roadster

The 440 centimetre long and 192 centimetre wide W 12 was made with the modern mono-coque construction method and featured upward swinging doors affixed to the front. Thanks to its lightweight construction the W 12, which was only 104 centimetres high, weighed about 1200 kilograms. In front it featured tyres sized 255/40 ZR 19, while in the rear the tyres sized 285/35 ZR 19. It reached a maximum speed of about 290 km/h.

Model	Volkswagen W 12 Roadster
Cubic capacity/cylinder	5584 cc/12 cylinder
hp/kW	420/309
Construction period	1997
Number of produced cars	unique piece

Zender Vision 3

In the late 1960s, Hans-Albrecht Zender began tuning serially produced cars. However, his enterprise in Mülheim-Kärlich also developed many interesting sports concept cars, for example, the "Vision 3". The 270 km/h-fast rear-wheel drive car was at the time among the fastest sports cars built in Germany.

Model	Zender Vision 3
Cubic capacity/cylinder	5584 cc/8 cylinder
hp/kW	300/220
Construction period	1987
Number of produced cars	2

AC Autocraft Cobra

As of 1986, on the British island close to Weybridge once again sports cars were produced that were entitled to carry the trade name Cobra. The manufacturing firm (AC Autocraft) used for the production of the aluminium bodywork, incidentally the old original mouldings, while a Ford engine appropriately powered the car. Nevertheless, in 1995 Autocraft once more ceased the production of the cars.

Model	AC Autocraft Cobra
Cubic capacity/cylinder	4942 cc/8 cylinder
hp/kW	224/165
Construction period	1986 – 1995
Number of produced cars	–

Aston Martin V8

When the second series of the DBS was produced in 1972, the car appeared with minor, but precise bodywork touch-ups. For example, the chrome-plated radiator grille as well as the "DB" logo (David Brown) now belonged to the past. David Brown had sold the stock interest of his enterprise to "Company Developments", and it was decided to introduce the model to the market as a "V8".

Model	Aston Martin V8
Cubic capacity/cylinder	5340 cc/8 cylinder
hp/kW	no information
Construction period	1973 – 1977
Number of produced cars	–

Aston Martin V8 Volante

In 1977, with the debut of the fifth series (the car was now called "V8 Vantage" and the convertible version "Volante") all exhaust technical problems seemed to have been solved. In spite of all that the V8 remained a toy for the affluent – its fuel consumption was around 24 litres of Super fuel for 100 kilometres. In return, it provided a driving feeling that was unparalleled.

Model	Aston Martin V8 Volante
Cubic capacity/cylinder	5340 cc/8 cylinder
hp/kW	no information
Construction period	1977–1982
Number of produced cars	–

Aston Martin V8

To be able to comply with the US exhaust regulations, Aston Martin launched within a short period of time the third and fourth series of the eight-cylinder coupé. Instead of a Bosch-Injection system, four Weber dual carburettors were used that were easier to adjust. The factory was very incommunicative about the output data by the way, any published information always consisted of mere estimates.

Model	Aston Martin V8
Cubic capacity/cylinder	5340 cc/8 cylinder
hp/kW	no information
Construction period	1977 – 1982
Number of produced cars	–

Aston Martin V8 Zagato Coupé

Actually, the V8 Zagato Coupé was intended to have a top speed of 300 km/h. However, this target was narrowly missed, as only 299 km/h could be measured. In the years 1987/88 a convertible spin off of this coupé was also available whose design rather needed some getting used to. A mere 37 units were produced and today this model belongs to the very rare models of the Aston Martin brand.

Model	Aston Martin V8 Zagato Coupé
Cubic capacity/cylinder	5340 cc/8 cylinder
hp/kW	no information
Construction period	1986 – 1988
Number of produced cars	52

Aston Martin Lagonda

In 1974, Aston Martin revived the noble Lagonda brand and presented a four-door sports saloon whose design was not to everybody's liking. With a height of only 130 centimetres, the square deluxe car (top speed of 236 km/h) was in many ways ahead of its time. It already featured a digital speedometer display, and the control buttons for light, wipers, etc. had to be only lightly touched thanks to its advanced sensor technology.

Model	Aston Martin Lagonda
Cubic capacity/cylinder	5340 cc/8 cylinder
hp/kW	no information
Construction period	1974 – 1989
Number of produced cars	645

Aston Martin Virage Coupé

In 1984 – the DBS or V8 coupé was still being built – Aston Martin commissioned five design studios to come up with a design of a successive model that should be launched no later than 1990 on the market. This resulted in the Virage, a design by the London "Royal College of Art". This design corresponded precisely to what Aston Martins had in mind and only had to be optimised in the wind tunnel.

Model	Aston Martin Virage Coupé
Cubic capacity/cylinder	5340 cc/8 cylinder
hp/kW	335/246
Construction period	1989 – 1994
Number of produced cars	–

Aston Martin Virage Volante

To be able to save costs, Aston Martin used the shortened chassis of the luxurious Lagonda sports saloon as the undercarriage for the new Virage. This chassis – a platform frame – was considered to be a technically mature construction that additionally was also suited for the open Virage Volante. Nevertheless, whether closed or open: both versions always provided maximum driving pleasure, as their maximum speed was approximately 254 km/h.

Model	Aston Martin Virage Volante
Cubic capacity/cylinder	5340 cc/8 cylinder
hp/kW	335/246
Construction period	1989 – 1994
Number of produced cars	–

Aston Martin Virage V8 Coupé

As of 1993, the Virage was alternatively available with an automatic gear. In addition, the manufacturer also launched another two longer versions with a larger wheel base, and offered so-called "conversion kits". These could not only be used to increase the output, but also the cubic capacity. All these measures were carried out at the plant for a surcharge of up to GBP 50.000 (!).

Model	Aston Martin Virage V8 Coupé
Cubic capacity/cylinder	5340 cc/8 cylinder
hp/kW	335/246
Construction period	1989 – 1994
Number of produced cars	–

Aston Martin Vantage

In 1992, at the Motor Show in Birmingham, Aston Martin after a long break once again exhibited a car with the model name Vantage. Basically a Virage, the Vantage offered some detailed optical as well as technical differences. Even though the Vantage was also fitted with a beefy V8 engine under its bonnet, the heavy car (1920 kilograms) reached only a maximum speed of 229 km/h.

Model	Aston Martin Vantage
Cubic capacity/cylinder	5340 cc/8 cylinder
hp/kW	550/404
Construction period	1992 – 1999
Number of produced cars	

Aston Martin Vantage Le Mans

For Aston Martin drivers who were in search of something special, the factory already offered in 1994 an interesting special edition – the Virage "Limited Edition Coupé" that was only produced ten times. Two years later, collectors could rejoice again, because on the occasion of the discontinuation of the Virage or Vantage-series, the last development stage consisted of the "Le-Mans" version, which was built exactly 40 times.

Model	Aston Martin Vantage Le Mans
Cubic capacity/cylinder	6374 cc/8 cylinder
hp/kW	600/441
Construction period	1996 – 1999
Number of produced cars	40

Aston Martin DB 7 Coupé

At Newport Pagnell in England, where the Aston Martin DB 7 was developed, the project was initially simply called NPX. The specialist press was very surprised when it was presented with the result of the project at the 1993 Geneva Motor Show. Following several years in which only "V-Cars" (Virage, Vantage and Volante) were built, finally, there was once again a "DB Model" – the new DB 7.

Model	Aston Martin DB7 Coupé
Cubic capacity/cylinder	3239 cc/6 cylinder
hp/kW	335/250
Construction period	1994 – 1999
Number of produced cars	–

Aston Martin DB 7 Volante

In 1994, when the production of the DB 7 commenced, the elegantly shaped car initially only enriched the streets as a coupé. The open version was introduced two years later. As opposed to the earlier company policy, the DB7 was not intended for production in an expensive and exclusive small series, but in larger volumes of produced cars. This concept paid off, and the DB 7, with a top speed of approximately 260 km/h was soon present in 29 countries.

Model	Aston Martin DB7 Volante
Cubic capacity/cylinder	3239 cc/6 cylinder
hp/kW	335/250
Construction period	1996 – 1999
Number of produced cars	–

Aston Martin DB 7 Vantage Coupé

In 1999, Aston Martin fans had reason to rejoice: under the bonnet of the further developed DB7 a silky-soft operating twelve-cylinder engine had been introduced. For nearly two years the engineers had been developing, and above all, testing the engine. Even during continuous long-term operation of 48 hours at the maximum speed (266 km/h), it did not exhibit any problems.

Model	Aston Martin DB7 Vantage Coupé
Cubic capacity/cylinder	5935 cc/12 cylinder
hp/kW	420/309
Construction period	1999–2003
Number of produced cars	–

Aston Martin DB 7 Vantage Volante

According to the statistics of the manufacturer, Aston Martin customers ordered more cars painted in green than in any other colour. In 1999, blue came in second place, while there were 200 colour shades altogether. According to a press release of the same year, Aston Martin used about 56,000 litres of paint only for the production of the DB 7 from 1994 to 1999!

Model	Aston Martin DB7 Vantage Volante
Cubic capacity/cylinder	5935 cc/12 cylinder
hp/kW	420/309
Construction period	1999–2003
Number of produced cars	–

Bentley Project Hunaudières

With the Hunaudières, Bentley introduced a sporty concept car, named after the famous straight stretch of the Le Mans Circuit. The body of the unique piece (4432 millimetres in length, 1998 millimetres wide) was coated with a pearlescent effect and was produced from a combination of light alloy and carbon fibre. The heart of the racing car was a 16-cylinder engine made of an aluminium alloy.

Model	Bentley Project Hunaudières
Cubic capacity/cylinder	8004 cc/12 cylinder
hp/kW	630/463
Construction period	1999
Number of produced cars	unique piece

Bristol 412

Bristol introduced the type 412 as a 2+2-seater deluxe convertible whose distinguishing feature was the unique construction of its folding roof. The bodywork of the 412 was made by Zagato in Milan and corresponded more to the Italian preferences than those of the British in the 1970s. For its propulsion, Bristol chose a powerful V8 engine by Chrysler.

Model	Bristol 412
Cubic capacity/cylinder	6556 cc/8 cylinder
hp/kW	210/154
Construction period	1976–1977
Number of produced cars	–

De Lorean DMC-2

Maybe the De Lorean was too far ahead of its time: this futuristically styled car made in Northern Ireland had a body made of rustproof high-grade steel. The shape of the 200 km/h-fast two-seater was designed by Italian star designer Giugiaro.

Model	De Lorean DMC-2
Cubic capacity/cylinder	2849 cc/8 cylinder
hp/kW	132/97
Construction period	1981–1982
Number of produced cars	8583

Jaguar E-Type Series 3

With the upgrade from six- to twelve-cylinder engines in 1971, Jaguar implemented a step that had long since been overdue. E-Type buyers appreciated this decision very much and accepted the fact that Jaguar at the same time mounted their favourite model exclusively on the chassis with the long wheel base. According to statistics, 87 percent of all E-Type Jaguars were exported while the remainder moved on British streets.

Model	Jaguar E-Type Series 3
Cubic capacity/cylinder	5354 cc/12 cylinder
hp/kW	276/202
Construction period	1971–1974
Number of produced cars	15,287

Jaguar E-Type Series 3

The Jaguar E-Type not only looked and performed excellently on boulevards and streets, it also held the licence to win which was in racing. The drivers who caused a great deal of attention at Brands Hatch and Silverstone included Graham Hill. The racing successes provided Jaguar with continually high sales figures until it ceased production in 1974.

Model	Jaguar E-Type Series 3
Cubic capacity/cylinder	5354 cc/12 cylinder
hp/kW	276/202
Construction period	1971–1974
Number of produced cars	15,287

Jaguar XJ 220 Prototype

It took until the Birmingham Motor Show of 1988 for Jaguar to once again present a pure-bred sports car like the E-Type. What's more, it even presented a super sports car – the Jaguar XJ 220. Its shape was futuristic – wide, low and fast. In addition, the prototype disposed of a four-wheel drive. Its aggressive silhouette was skilfully emphasised by the choice of extra-wide tyres.

Model	Jaguar XJ 220 Prototype
Cubic capacity/cylinder	6222 cc/12 cylinder
hp/kW	500/373
Construction period	1988
Number of produced cars	2

Jaguar XJ 220

After the successful debut of the XJ-220 prototype, Jaguar decided to introduce this super sports car to the market in a slightly modified form, in a small volume series. However, as opposed to earlier plans, the car was not equipped with four-wheel drive technology, and instead of the twelve-cylinder engine the 320 km/h-fast serial version was instead fitted with a highly competitive V6 turbo engine.

Model	Jaguar XJ 220
Cubic capacity/cylinder	3498 cc/6 cylinder
hp/kW	549/404
Construction period	1992 – 1994
Number of produced cars	280

Jaguar XJ-S Coupé

When Jaguar constructed the XJ-S coupé, it had intended for this car to follow in the footsteps of the E-Type. Although a V12 engine powered the 240 km/h-fast XJ-S, it unfortunately never achieved the sporty image of the E-Type. The sales figures could neither be improved by an inexpensive alternative with a six-cylinder-engine, nor by the luxurious convertible version.

Model	Jaguar XJ-S Coupé
Cubic capacity/cylinder	5343 cc/12 cylinder
hp/kW	287/210
Construction period	1974 – 1981
Number of produced cars	–

Jaguar XJ-SC 5.3

While the introduction of the XJ-SC in 1983 meant that after a long break, Jaguar once again featured an (almost) open car in its program, the removable roof concept did not seem to be a felicitous solution. It was not until 1988 that a "real" fully convertible car based on the XJ-SC went into production. The heavy car that was very thirsty for fuel was powered by a V 12 engine and reached a maximum speed of 230 km/h.

Model	Jaguar XJ-SC 5.3
Cubic capacity/cylinder	5343 cc/12 cylinder
hp/kW	295/217
Construction period	1988 – 1992
Number of produced cars	–

Jaguar XJ 6 Compressor

The presentation of the first generation of the XJ saloon class took place in 1968. In retrospect it can be said that this was the beginning of a new age for Jaguar. The styling of the timelessly beautiful saloon, of which there also existed and still exist today sportive styled variations, was still created to a great extent by Sir William Lyons, the founder and head of Jaguar at that time.

Model	Jaguar XJ 6 Compressor
Cubic capacity/cylinder	3980 cc/6 cylinder
hp/kW	320/235
Construction period	1994 – 1997
Number of produced cars	–

Jaguar XJR 4.0 l

In the early 1990s, the XJ series was already produced in the fifth generation, and Jaguar's new owner Ford invested nearly GBP 200 million into development efforts. In 1994, the XJ was available alternatively with a 3.2- or 4.0-litre six-cylinder engine. However, the highlight of this model year was the XJR equipped with a four-litre-compressor engine with a long wheel base.

Model	Jaguar XJR 4.0 l
Cubic capacity/cylinder	3980 cc/6 cylinder
hp/kW	320/235
Construction period	1994 – 1997
Number of produced cars	–

Jaguar XK 8

The history of the XK class goes back to 1948. It was distinguished by the claim to build sporty vehicles with exciting designs. Since 1996, Jaguar once again featured an XK series in its programme, when the new XK 8 celebrated its première at the Paris International Motor Show. Jaguar had once again created a sports car whose high-calibre technology lived up to the promise of its classy shape.

Model	Jaguar XK 8
Cubic capacity/cylinder	3996 cc/8 cylinder
hp/kW	284/209
Construction period	1996 – 2001
Number of produced cars	–

Jaguar XK 8

Jaguar introduced the XK series in 1996 in a convertible and a coupé version. The novelty of these cars was the fact that they were equipped with V8 engines. This certainly took the driving pleasure into a new dimension, although the manufacturer limited the maximum speed to no more than 250 km/h. Incidentally, in the model year 2002, the XK was subjected to the first styling makeovers.

Model	Jaguar XK 8
Cubic capacity/cylinder	3996 cc/8 cylinder
hp/kW	284/209
Construction period	1996 – 2001
Number of produced cars	–

Jaguar XK 180

With the concept car study XK 180 shown in 1998 at the Paris Motor Show, Jaguar commemorated the 50th anniversary of the market launch of the first XK sports car (XK 120). From the outset, the beefy XK 180 with its main headlights concealed under Plexiglas was not planned for serial production. For a car show in the USA, the manufacturer produced another unit in 1999.

Model	Jaguar XK 180
Cubic capacity/cylinder	no information
hp/kW	no information
Construction period	1998 – 1999
Number of produced cars	2

Jaguar F-Type Concept Roadster

With the F-Type Concept, the most compact Jaguar sports car created in over 40 years, the engineers continued the development of the legendary E-Type and created a roadster with which Jaguar could access a new market segment. By presenting the model at international motor shows, Jaguar wanted to test the reaction of the audience to this design concept.

Model	Jaguar F-Type Concept Roadster
Cubic capacity/cylinder	no information
hp/kW	no information
Construction period	1999/2000
Number of produced cars	unique piece

Jaguar F-Type Concept Roadster

The F-Type was developed by a team of designers and engineers under the direction of Keith Helfet who was already responsible for the XK-180 concept and the XJ-220 sports car. Because neither technical nor style restrictions were imposed on the team, its members were able to create a car with an absolutely independent character.

Model	Jaguar F-Type Concept Roadster
Cubic capacity/cylinder	no information
hp/kW	no information
Construction period	1999/2000
Number of produced cars	unique piece

Jensen S V8

The British sports car brand Jensen also hoped in the middle of the 1990s for a come-back. Under the direction of Graham Morris, a former Rolls Royce employee, the enterprise was newly established once again. When the shiny new Jensen Roadster became available in 1999, the makers had supposedly already received 200 orders. However, unfortunately there was only enough money for the construction of 24 units.

Model	Jensen S V8
Cubic capacity/cylinder	4600 cc/8 cylinder
hp/kW	316/232
Construction period	1998 – 2002
Number of produced cars	24

Lotus Elite

Experts certainly appreciated the hyper-modern design of the Elite introduced in 1973. Conservative customers, however, were mostly shocked by this modern car with the straight hatchback rear. There were also many novelties to admire under its bonnet: this was where a two-litre engine with 16-valve technology propelled the low plastic car to a maximum speed of 195 km/h.

Model	Lotus Elite
Cubic capacity/cylinder	1974 cc/4 cylinder
hp/kW	160/118
Construction period	1973 – 1980
Number of produced cars	2398

Lotus Esprit S 1

In 1974, Lotus commissioned Giorgio Giugiaro to develop a design study, which should form the base of a wedge-shaped sports car. The engine unit, a four-cylinder with 16-valve technology, should be placed vertically before the rear axle and provide the car with a maximum speed of at least 210 km/h. This goal was achieved and the sporty wedge of a car went into production in 1975 under the model name Esprit.

Model	Lotus Esprit S 1
Cubic capacity/cylinder	1974 cc/4 cylinder
hp/kW	160/117
Construction period	1975 – 1980
Number of produced cars	718

Lotus Esprit V8

With the timeless design of the Esprit, Lotus launched a model that corresponded exactly to the wishes of its customers. Due to excellent sales figures, the model range developed quickly. The last stage of development included the Esprit V8 with a maximum speed of 282 km/h. The fuel bill was accordingly high – the car's average consumption was around 20 litres for each 100 kilometres.

Model	Lotus Esprit V8
Cubic capacity/cylinder	3500 cc/8 cylinder
hp/kW	354/260
Construction period	1997 – 2001
Number of produced cars	–

Lotus Elise

In the middle of the 1990s, the Lotus engineers developed the Elise, a car that according to its press information was "great fun to drive". The fun mobile was based on an aluminium frame and weighed just 690 kilograms. In connection with its 120-hp engine, a maximum speed of 202 km/h could be reached by it.

Model	Lotus Elise
Cubic capacity/cylinder	1796 cc/4 cylinder
hp/kW	120/88
Construction period	1997
Number of produced cars	–

Lotus Elise 111 S

In the by now ten-year history of the rustic Lotus Elise various changes have already been introduced. In 1997, an optional hard top became available for the low two-seater, two years later the mid-engine sports car received a considerable boost to its output achievement and broader tyres. The first minor face-lift took place in autumn of the year 2000.

Model	Lotus Elise 111 S
Cubic capacity/cylinder	1796 cc/4 cylinder
hp/kW	145/107
Construction period	1999 – 2000
Number of produced cars	–

Lotus Elise 340 R

The irresistible look of the 340 R and its minimalist bodywork (no roof!) make this car that costs approximately EUR 52,000 an uncompromising sports car for true car enthusiasts. The special model was based on a prize-winning design concept car, which Lotus introduced in 1998 at the Birmingham Motor Show.

Model	Lotus Elise 340 R
Cubic capacity/cylinder	1796 cc/4 cylinder
hp/kW	177/130
Construction period	1999 – 2000
Number of produced cars	340

MG B GT V8

In its original design as a compact hatchback coupé with a wide opening boot lid, the MG GT was a rather practical and sporty car at the same time. It was equipped with a four-cylinder engine and remained in production from 1965 to 1980. Its offshoot, the far more powerful MG B GT V8, hardly sold at all despite its V8 engine and powerful output.

Model	MG B GT V8
Cubic capacity/cylinder	3528 cc/8 cylinder
hp/kW	137/101
Construction period	1973 – 1976
Number of produced cars	approx. 2590

MG Type B

When the shape of the MG A in the early 1960s did not completely correspond to the contemporary taste anymore, MG introduced to the market the new MG B as its successor. In 1965 a closed variation was juxtaposed to the perky roadster in the shape of a hatchback coupé. The MG B sold exceptionally well – unfortunately, the model was provided as of 1974 with bulky plastic fenders.

Model	MG Type B
Cubic capacity/cylinder	1798 cc/4 cylinder
hp/kW	95/70
Construction period	1974–1980
Number of produced cars	

MG GT V8

The GT V8 provided an especially attractive combination of sportiness and driving pleasure. In the beginning of the 1990s, the easily manageable two-seater (233 centimetres wheel base) distinguished itself positively from other convertible cars. A roll bar that could interfere with the overall harmonious silhouette was foreign to the well-proportioned Briton.

Model	MG GT V8
Cubic capacity/cylinder	3946 cc/8 cylinder
hp/kW	190/140
Construction period	1993 – 1996
Number of produced cars	approx. 2300

MG F

When the MG F debuted in 1995, there was finally again a roadster that continued the tradition of the typically British sports car. With a 1.8 litre four cylinder engine the MG F offered affordable pleasure. Due to its rather short wheel base (237.5 centimetres) the car, similar to its ancestors from the 1950s, was conceived as a compact two-seater.

Model	MG F
Cubic capacity/cylinder	1796 cc/4 cylinder
hp/kW	150/110
Construction period	1995 – 2001
Number of produced cars	approx. 77.500

TVR S4C

In addition to the domestic market, the sales in the US market play an especially important role for TVR. One more reason why the company has favoured for a long time already the installation of engines with high cubic capacities. As an alternative to the highest output level, the TVR V8S Roadster was also offered as a slightly less agile version with a six-cylinder engine (S4C).

Model	TVR S4C
Cubic capacity/cylinder	2933 cc/6 cylinder
hp/kW	175/129
Construction period	1991 – 1997
Number of produced cars	–

Bugatti EB 18/3 Chiron

When the Volkswagen AG bought the name Bugatti in 1995, it commissioned Italdesign with the design of the Bugatti EB 218. This elegant touring car was the absolute star of the 1999 Geneva Salon. Some variations of this model followed, including the Bugatti 18/3 Chiron. Nevertheless, a final solution for the serially produced model was found only with the Bugatti Veyron built since 2005.

Model	Bugatti EB 18/3 Chiron
Cubic capacity/cylinder	6255 cc/18 cylinder
hp/kW	555/408
Construction period	1999
Number of produced cars	unique piece

Bugatti EB 110 SS

In 1987 the Italian entrepreneur Romano Artiolo bought the rights to the name of the former deluxe brand Bugatti. Two years later his factory in Modena once again produced sports cars that carried the trade name Bugatti – the models EB 110 GT and EB 110 SS. In 1995, the Bugatti Automobili Spa ceased its operations, and the rights changed their owner once again, this time in favour of the Volkswagen AG.

Model	Bugatti EB 110 SS
Cubic capacity/cylinder	3500 cc/12 cylinder
hp/kW	600/441
Construction period	1992 – 1995
Number of produced cars	32

Citroen SM

It was a perfect surprise – in 1970, Citroen presented the model SM, which was not only a technical miracle, but also the world's fastest front-wheel drive vehicle of the time! This car was a co-operative effort between Citroen and Maserati. While Citroen was responsible for the ingenious hydropneumatic suspension system, Maserati contributed the V6 engine featuring four overhead camshafts.

Model	Citroen SM
Cubic capacity/cylinder	2670 cc/6 cylinder
hp/kW	170/125
Construction period	1970 – 1975
Number of produced cars	12.920

Matra Bagheera

If there ever was a car in which three persons could sit comfortably side by side in front, it was the Matra Bagheera. Simca contributed the technology of the plastic car, because Matra belonged to the Chrysler-Simca sales group. Matra offered the Bagheera with two output levels. Above all, the latest model equipped with a 1.5-litre engine promised pure driving pleasure – a speed of 185 km/h was always available.

Model	Matra Bagheera
Cubic capacity/cylinder	1294 cc/4 cylinder
hp/kW	85/62
Construction period	1973 – 1980
Number of produced cars	47.002

Renault Alpine A 310

In 1971, Alpine presented a successor of the A 110 – the slightly larger A 310. Its bodywork was also made of plastic. The A 310 was powered by an engine positioned in the rear which normally was used for the Renault 16 TX. Unfortunately, the engine that was modified to provide more sportiness handled rather roughly, which was why Alpine preferred to install a V6 engine as of 1977.

Model	Alpine A 310
Cubic capacity/cylinder	1605 cc/4 cylinder
hp/kW	127/93
Construction period	1971 – 1976
Number of produced cars	2340

Renault Alpine A 310

Starting in 1977, the Alpine A 310 received instead of its four-cylinder engine a stronger six-cylinder which had already been tested in the Renault 30. Optically, the newly designed headlights, three-bolt wheels and the widened track mainly distinguished the six-cylinder Alpine. The interior profited from so-called pilot's seats with adjustable lateral support.

Model	Alpine A 310
Cubic capacity/cylinder	2664 cc/6 cylinder
hp/kW	150/110
Construction period	1977 – 1981
Number of produced cars	approx. 9200

Renault Alpine A 310

As part of the model development, the A 310 was only slightly modified. In 1979 a five-speed gearbox became available upon request, while the model year 1981 brought new wide plastic fenders and bigger wheels. To accommodate the wider tyres, the fender openings were modified – a cosmetic measure that provided the car with an even sportier look.

Model	Alpine A 310
Cubic capacity/cylinder	2664 cc/6 cylinder
hp/kW	150/110
Construction period	1981 – 1984
Number of produced cars	approx. 9200

Renault Alpine A 310

Although from 1981 Renault owned the majority of Alpine shares, and the car was therefore called Renault Alpine A 310, the fans of the 225 km/h-fast sports car continued to call it the "Alpine". To them, the name of the best-seller played a subordinate role. After nearly 10,000 built units, the first A 310 generation made room in 1985 for the second construction series.

Model	Alpine A 310
Cubic capacity/cylinder	2664 cc/6 cylinder
hp/kW	150/110
Construction period	1981 – 1984
Number of produced cars	approx. 9200

Renault Alpine V6 GT Turbo

Renault introduced the V6 GT Turbo as a successor of the V6 GT model that had been built in 1985/86. With this car, the plant offered an especially sporty model (243 km/h maximum speed) with its own character and the amenities of a high class vehicle. The principle of the propulsion concept remained unchanged, and the engine continued to be installed in the rear.

Model	Alpine V6 GT Turbo
Cubic capacity/cylinder	2458 cc/6 cylinder
hp/kW	185/136
Construction period	1986 – 1991
Number of produced cars	approx. 4980

Renault Alpine V6 GT Turbo

For the production of the car's plastic unibody, Alpine had developed a process in which a supportive steel frame was strengthened by and covered in a polyester material. The design engineers paid particular attention to the air resistance – thanks to the smooth silhouette and the wainscoted underbody, the Alpine exhibited a drag coefficient of a mere 0.30 c_d.

Model	Alpine V6 GT Turbo
Cubic capacity/cylinder	2458 cc/6 cylinder
hp/kW	200/147
Construction period	1986 – 1991
Number of produced cars	325

Renault Alpine A 610 Turbo

The heart of the A 610 was a Turbo engine made of light alloy with three litres of cubic capacity. A Garrett exhaust turbocharger served the modern injection system, initiating a very early torque. This was the only way in which the high output levels of 250 hp could be achieved. The exhaust system made of stainless steel already contained a three-road catalytic converter.

Model	Alpine A 610 Turbo
Cubic capacity/cylinder	2975 cc/6 cylinder
hp/kW	250/184
Construction period	1991 – 1994
Number of produced cars	820

Renault Alpine A 610 Turbo

The highly competitive six-cylinder engine, the relatively low weight of the car (1420 kilograms) and an aerodynamic body combined to provide the best prerequisites for the exceptional driving characteristics of the A 610. For example, it accelerated from 0 to 100 km/h in only 5.9 seconds – while its maximum speed amounted to 265 km/h.

Model	Alpine A 610 Turbo
Cubic capacity/cylinder	2975 cc/6 cylinder
hp/kW	250/184
Construction period	1991 – 1994
Number of produced cars	820

Renault Alpine A 610 Turbo

Beyond any doubt, the A 610 could be considered one of the most beautiful French sports cars. It was therefore not surprising that the low coupé with the sleepy eyes-headlights was also in demand in the USA.

Unfortunately, the model intended for the US market had to be equipped with broader fenders, since this was legally required there.

Model	Alpine A 610 Turbo
Cubic capacity/cylinder	2975 cc/6 cylinder
hp/kW	250/184
Construction period	1991 – 1994
Number of produced cars	820

Renault 5 Turbo

To be able to take part in the 1980 Rally World Championship, Renault introduced the R5 Turbo, a homologation car whose tuned up engine was not mounted in front, but rather positioned in place of the back bench. In addition, the car that was thus mutated into a two-seater received a newly contrived chassis and was equipped with extremely wide tyres.

Model	Renault 5 Turbo
Cubic capacity/cylinder	1397 cc/4 cylinder
hp/kW	160/118
Construction period	1980 – 1986
Number of produced cars	approx. 4000

Renault Spider

"Conceived for racing – adjusted to street use", read the title of a press release published by Renault on the occasion of the debut of the Spider. This was because the Spider was the result of a great passion of the manufacturer – the racing sport. As a result, the owners of such a "Duoposto", could drive it not only on the street, but also complete rounds on the racetrack with it.

Model	Renault Spider
Cubic capacity/cylinder	1998 cc/4 cylinder
hp/kW	147/108
Construction period	1996 – 1999
Number of produced cars	1640

Renault Spider

To allow customers to fully and individually enjoy to the car's driving pleasure, Renault introduced the Spider in two versions. In addition to the model with a small air deflector, there was another variation with a full windshield. In terms of technology, both models were identical and had a maximum speed of about 215 km/h (acceleration from 0 to 100 km/h in 6.9 seconds).

Model	Renault Spider
Cubic capacity/cylinder	1998 cc/4 cylinder
hp/kW	147/108
Construction period	1996 – 1999
Number of produced cars	1640

Renault Spider Sport

The body of the Spiders consisted – with the exception of the doors – of only three parts: the front and rear bonnets, as well as the middle section. All three parts were manufactured of light composite materials and could be completely removed. For the stable floor of the car, the design engineers opted for honeycomb-shaped, two centimetre thick Polypropylene.

Model	Renault Spider
Cubic capacity/cylinder	1998 cc/4 cylinder
hp/kW	147/108
Construction period	1996 – 1999
Number of produced cars	1640

Renault Spider Sport

Originally, the objective of the development of the Spider was for use in racetracks only. However, since it was street legal without any restrictions and there was great positive resonance among fans. As a pure-bred roadster, the Spider immediately advanced to the status of an absolute and scarce cult object. Keeping in line with its sportliness, the interior was rather sparsely equipped.

Model	Renault Spider Sport
Cubic capacity/cylinder	1998 cc/4 cylinder
hp/kW	147/108
Construction period	1996 – 1999
Number of produced cars	1640

Renault Clio Sport V6 24 V

The Clio Sport V6 24 V celebrated its première as a concept car at the 1998 Paris Motor Show. Actually, Renault wanted to use it only to test the reactions of the audience. However, since the car awakened recollections of the days of the Renault 5 Turbo, Renault was left with no option but to develop the Clio concept car and launch it in a small series.

Model	Renault Clio Sport V6 24 V
Cubic capacity/cylinder	2946 cc/6 cylinder
hp/kW	226/166
Construction period	1999 – 2002
Number of produced cars	1647

Venturi Atlantique 300

In 1984, former Renault Alpine employees founded the Venturi company. Following several failed attempts, they achieved their goal of developing and establishing a sports car only ten years later with the Atlantique model. The Atlantique, a mid-engine sports car with plastic bodywork, cost about EUR 80,000. Because a Venturi was always fully equipped, there was no (!) surcharge list.

Model	Venturi Atlantique 300
Cubic capacity/cylinder	2975 cc/6 cylinder
hp/kW	310/228
Construction period	1995 – 2001
Number of produced cars	–

Venturi Atlantique 300

The Atlantique 300 may claim to have been most successful of all Venturi cars. The French manufacturer equipped it with the so-called Euro V6 light alloy engine. However, Venturi tuned this engine, and offered it in several output versions. In the 281-hp version, it provided the Atlantique with a substantial maximum speed of 280 km/h.

Model	Venturi Atlantique 300
Cubic capacity/cylinder	2975 cc/6 cylinder
hp/kW	281/207
Construction period	1995 – 2001
Number of produced cars	–

Alfa Romeo Montreal

In 1970, Alfa Romeo finally once again introduced a surprising eight-cylinder engine car, the Montreal. Strictly speaking, it was the advancement of a mid-engine model already shown in 1967, which had now been reconstructed into a street sports car with the engine placed in front. Unfortunately, this change made the Montreal a little heavy and ungainly.

Model	Alfa Romeo Montreal
Cubic capacity/cylinder	2593 cc/8 cylinder
hp/kW	200/147
Construction period	1970 – 1975
Number of produced cars	3925

Alfa Romeo Alfetta GT

In 1972, Alfa Romeo introduced the Alfetta, a modern four-door saloon, to the general public. Production of its matching counterpart, a two-door sports coupé, commenced two years later under the model name Alfetta GT. Buyers could choose among a 1.6-, 1.8- and a two-litre four-cylinder engine. With such an engine, the coupé reached a maximum speed ranging from 180 to 210 km/h.

Model	Alfa Romeo Alfetta GT
Cubic capacity/cylinder	1570 cc/4 cylinder
hp/kW	109/80
Construction period	1974 – 1976
Number of produced cars	–

Alfa Romeo Alfetta GTV 2.5

To be able to fulfil the customers' desire for more performance, the Alfetta GT was offered from 1976 with a 122 hp, two-litre four-cylinder engine. For the model of the year 1980 they came up with a further increase – now the current version of the coupé was given the name Alfetta GTV 6 2.5 and featured, on account of a six-cylinder engine, an elevation on the hood.

Model	Alfa Romeo Alfetta GTV 2.5
Cubic capacity/cylinder	2492 cc/6 cylinder
hp/kW	160/118
Construction period	1980 – 1986
Number of produced cars	–

Alfa Romeo Alfetta GTV 2.5

With the Alfetta GTV 2.5 Alfa Romeo carried a prestigious coupé in its program that completely fulfilled the desires of drivers with racing ambitions. It perkily reached 215 km/h, had a precisely dimensioned chassis and an aerodynamic bodywork (c_d value 0.39). The basic equipment included height-adjustable front seats as well as a classical wooden steering wheel.

Model	Alfa Romeo Alfetta GTV 2.5
Cubic capacity/cylinder	2492 cc/6 cylinder
hp/kW	160/118
Construction period	1980 – 1986
Number of produced cars	–

Alfa Romeo R.Z.

The fascinating design, which made the 230 km/h-fast R.Z. an object of desire, originated once again on the drawing board of the master coachmaker Zagato. The construction made of glass fibre was manually fitted on the slightly modified foundation of Alfa Romeo 75. However, with a list price of DM 130.000, the manufacturer also asked customers to pay a considerable price for this successful synthesis.

Model	Alfa Romeo R.Z.
Cubic capacity/cylinder	2959 cc/6 cylinder
hp/kW	210/154
Construction period	1989 – 1994
Number of produced cars	1000

Alfa Romeo R.Z.

The precursor of the Alfa Romeo R.Z. (ES Concept) was already shown in 1986. Now, three years later, this extremely sporty car actually became available. Initially, the plant introduced it as a coupé version (Alfa Romeo S.Z.), while the roadster variation shown here (Alfa Romeo R.Z.) followed a little later. A total of 1000 cars were built, including 241 roadsters.

Model	Alfa Romeo R.Z.
Cubic capacity/cylinder	2959 cc/6 cylinder
hp/kW	210/154
Construction period	1989 – 1994
Number of produced cars	1000

Alfa Romeo Proteo 164

With the concept car 164 Proteo shown in 1991 (maximum speed 250 km/h), Alfa Romeo once again presented an independent sports car design, whose serial production was implemented a short time afterwards (Spider and GTV as of 1994). The Proteo, featuring a wedge-shaped and very dynamic design, was based on the shortened underbody of the big Alfa Romeo saloon (Type 164) and equipped with a V6 engine.

Model	Alfa Romeo Proteo 164
Cubic capacity/cylinder	2959 cc/6 cylinder
hp/kW	260/191
Construction period	1991
Number of produced cars	unique piece

Alfa Romeo Nuvola

Fortunately, the Alfa Romeo Nuvola did not belong to the concept cars that disappeared into obscurity for good after their debut. The 280 km/h-fast Nuvola concept car (front track width 1577 millimetres, rear track width 1602 millimetres) was refined once again in 2004 and this upgraded variation served as the base of another show car, the 8C Competizione.

Model	Alfa Romeo Nuvola
Cubic capacity/cylinder	2959 cc/6 cylinder
hp/kW	304/224
Construction period	1996
Number of produced cars	unique piece

Alfa Romeo Spider

When Alfa Romeo produced the final version of the legendary Spider, after a long wait, the fans finally received the car that they had actually desired right from the start. Finally, the shape was right, and there were neither ugly spoilers nor rubber extensions. The variation was available with a 1.6- or 2.0-litre engine, while the maximum speed was between 180 and 195 km/h.

Model	Alfa Romeo Spider
Cubic capacity/cylinder	1962 cc/4 cylinder
hp/kW	120/88
Construction period	1990 – 1994
Number of produced cars	–

Bizzarrini BZ-2001

After Bizzarrini gave up automobile construction in 1969 due to financial reasons, since the 1990s there were several attempts to reanimate the brand once again. For example in 1994, a 300 km/h-fast spider called BZ-2001 awakened the hopes of the sports car scene. However, as frequently true in automobile history, this project also got stuck in the beginning stages.

Model	Bizzarrini BZ-2001
Cubic capacity/cylinder	4942 cc/12 cylinder
hp/kW	360/265
Construction period	1994
Number of produced cars	unique piece

Cizeta Moroder V 16 T

Giorgio Moroder, a composer by profession, together with the Italian engineer Claudio Zampolli developed in the middle of the 1980s their own super sports car, the Moroder V16 T. However, following the debut of the car, they received merely ten orders. After all, not everyone had USD 600,000 to spare for a car. The most prominent Moroder owner is, incidentally, the Sultan of Brunei: he drives two of these race cars.

Model	Cizeta Moroder V 16 T
Cubic capacity/cylinder	5995 cc/16 cylinder
hp/kW	540/397
Construction period	1988 – 1993
Number of produced cars	–

Dino 308 GT 4

As opposed to its six-cylinder predecessor Dino 246 GT, the Dino 308 GT 4, designed by Bertone, was given an eight-cylinder engine. From 1975 to 1980, Ferrari additionally released the smaller 208 GT 4 (cubic capacity 1991 cubic centimetres, 170 hp). This model version, which was built 840 times, exclusively served the Italian market.

Model	Dino 308 GT 4
Cubic capacity/cylinder	2926 cc/8 cylinder
hp/kW	236/174
Construction period	1973 – 1980
Number of produced cars	2826

De Tomaso Panthera

The formerly Argentinian racing driver Alejandro de Tomaso, who was highly successful for Maserati and Osca, founded his own car factory in 1959. However, the initially built Monoposto racers barely earned any money. The brand only became really famous after the debut of the best-seller Panthera, which was offered in several output levels and versions.

Model	De Tomaso Panthera
Cubic capacity/cylinder	5796 cc/8 cylinder
hp/kW	330/243
Construction period	1970 – 1996
Number of produced cars	approx. 9000

Ferrari 365 GTC/4

Although the design of 365 GTC/4 was designed by Pininfarina, in the opinion of many car lovers this car was nothing more than a carelessly fast design. Granted, the look of the GTC took some getting used to, but one had to drive this car first to really appreciate its qualities. There were also supposedly Daytona owners who owned a 365 GTC/4 because it caused less problems.

Model	Ferrari 365 GTC/4
Cubic capacity/cylinder	4390 cc/12 cylinder
hp/kW	330/242
Construction period	1971 – 1972
Number of produced cars	480

Ferrari 365 GT/4 BB

The Berlinetta Boxer (BB) was the first street sports car by Ferrari with a mid-engine system. The 4.4-litre twelve-cylinder engine featured four overhead camshafts, which – another novelty – were not moved by a chain, but with the help of two toothed drive belts. The car's road performance: acceleration from 0 to 100 km/h in six seconds with a maximum speed of 278 km/h.

Model	Ferrari 365 GT/4 BB
Cubic capacity/cylinder	4391 cc/12 cylinder
hp/kW	351/258
Construction period	1971 – 1976
Number of produced cars	387

Ferrari 512 BB

As part of the further advancement and model development, the 365 GT/4 BB was followed in late 1976 by the 512 BB. For this car the model name no longer referred to the cubic capacity of a single cylinder. Rather, the "five" stood for a total of five litres of cubic capacity and the "12" signified twelve cylinders. The versions built from 1981 with electronic fuel injection carried the name 512 BBi.

Model	Ferrari 512 BB
Cubic capacity/cylinder	4943 cc/12 cylinder
hp/kW	360/265
Construction period	1976 – 1981
Number of produced cars	929

Ferrari Testarossa

The design of the Testarossa, which needed some getting used to, did not match everybody's taste. However, the curved slits along the sides fulfilled their purpose – they supplied a substantial amount of fresh air to the radiators that were positioned before the rear wheels. The twelve-cylinder flat engine provided the Testarossa with a maximum speed of 293 km/h.

Model	Ferrari Testarossa
Cubic capacity/cylinder	4942 cc/12 cylinder
hp/kW	390/287
Construction period	1984 – 1996
Number of produced cars	9937

Ferrari 308 GTB

The construction model 308, which 30 years ago was rather frowned upon as a "common-folk's Ferrari", has like all Ferraris, in the mean-time achieved collector's status. It remained in production for nearly 15 years and was available with different body variations. The 308 GTB (or 328 GTB as of 1985) was the over-all best selling Ferrari to date. Its maximum speed was according to engine power between 242 km/h and 267 km/h.

Model	Ferrari 308 GTB
Cubic capacity/cylinder	2926 cc/8 cylinder
hp/kW	227/167
Construction period	1975 – 1989
Number of produced cars	21.700 (all series)

Ferrari 288 GTO

At first glance, the GTO looks rather like an expanded 308 GTB. However, the basic data of this agile car with its eight-cylinder engine (305 km/h maximum speed) speak for themselves. Its lengthways rather than transversely installed double turbo engine had an output of 400 hp and turned with 7000 rev./min. The GTO was available in only one colour: Ferrari Red!

Model	Ferrari 288 GTO
Cubic capacity/cylinder	2855 cc/8 cylinder
hp/kW	400/294
Construction period	1984 – 1986
Number of produced cars	272

Ferrari F 40

The F 40, conceived as the successor of the GTO, enriched the sports car scene conveniently in time for the company's 40th anniversary. The speedometer needle of this last sports car only stopped at 324 km/h, while its acceleration from a standstill to 100 km/h took only 4.6 seconds. While as a new car the F 40 cost DM 260.000, this object of desire today fetches multiples of this amount.

Model	Ferrari F 40
Cubic capacity/cylinder	2936 cc/8 cylinder
hp/kW	478/351
Construction period	1987 – 1992
Number of produced cars	1311

Ferrari 348 TB/TS

The Ferrari 348, successor of the 308/328 series, optically resembled the style of the big Testarossa. The styling of the 348 was actually based on a design by Pininfarina – in addition to the closed coupé variation (Berlinetta TB) the sports car with a top speed of 275 km/h was also available with a removable middle roof section (Spider TS).

Model	Ferrari 348 TB/TS
Cubic capacity/cylinder	3405 cc/8 cylinder
hp/kW	300/221
Construction period	1989 – 1994
Number of produced cars	9485

Ferrari 456 GT

The Ferrari 456 that continued the tradition of the brand in the area of 2 + 2-seater cars, provided the output strength of a V12 engine coupled with the comfort of a GT vehicle. To provide more space in the interior the engine was placed in front and the transmission in the rear. For the purpose of reducing the vehicle weight, Ferrari constructed the hood of carbon fibre material.

Model	Ferrari 456 GT
Cubic capacity/cylinder	5473 cc/12 cylinder
hp/kW	442/325
Construction period	1992 – 2002
Number of produced cars	3287

Ferrari F 355 Spider

50 journalists were the first persons allowed to critically examine the Ferrari 355 in May 1994. The 355, which was available in the versions Berlinetta, GTS, Spider and Challenge, was especially compelling with its fiery eight-cylinder engine. Its strength was transmitted via a six-step gear transmission to the rear axle and provided a maximum speed of 295 km/h.

Model	Ferrari F 355 Spider
Cubic capacity/cylinder	3496 cc/8 cylinder
hp/kW	380/279
Construction period	1994 – 1998
Number of produced cars	11.258

Fiat X 1/9

At first the X 1/9 was only a concept car designed by Bertone. In 1972 however, the model went into production to the great joy of many enthusiasts. A good decision, because the X 1/9 with a top speed of 175 km/h of sold well. Unfortunately, Fiat ceased its production in 1982. Based on the great demand, Bertone continued the production on its own – the models built until 1989 were marketed under the name Bertone X 1/9.

Model	Fiat X 1/9
Cubic capacity/cylinder	1290 cc/4 cylinder
hp/kW	75/55
Construction period	1972 – 1982
Number of produced cars	approx 180.000

Ferrari 550 Maranello

With the 550 Maranello, Ferrari presented an aesthetically perfect car for customers who greatly valued comfort and convenience in addition to sporty road performance. Accordingly, the finest Conolly leather dominated the noble interior of this two-seater, and air-conditioning as well as a Hi-fi system belonged to its basic equipment.

Model	Ferrari 550 Maranello
Cubic capacity/cylinder	5474 cc/12 cylinder
hp/kW	485/357
Construction period	1996 – 2005
Number of produced cars	–

Lamborghini Miura Jota

Thanks to their powerful output, the street versions of the Miura built until 1971 achieved top speeds of approximately 275 km/h. Not a bad value, yet, this was by far not the maximum performance the concept of this car could provide. Among the roughly 800 manufactured cars there were quite a few special models with up to 440 hp. The carmaker also manufactured another unique piece with a sporty racing character, the Miura Jota, shown here.

Model	Lamborghini Miura Jota
Cubic capacity/cylinder	3929 cc/12 cylinder
hp/kW	401/297
Construction period	1970
Number of produced cars	–

Fiat 124 Spider Volumex

Driving a Fiat Spider in an extremely sporty way was only possible with the Spider Volumex. As opposed to the other Spider models, a so-called Roots supercharger provided plenty of output under its bonnet. Originally equipped with light alloy wheels and discreetly enlarged fenders, the Volumex reached a maximum speed of 205 km/h.

Model	Fiat 124 Spider Volumex
Cubic capacity/cylinder	1995 cc/4 cylinder
hp/kW	135/99
Construction period	1984 – 1985
Number of produced cars	–

Lamborghini Jarama

The Jarama, a really fast and agile car, was created as of 1970 to follow in the footsteps of the Islero. Styled totally differently from its outdated predecessor, the Jarama profited from numerous technical innovations. It was based on a completely newly developed underbody concept (self-supporting steel platform frame) which was derived from the Espada.

Model	Lamborghini Jamara
Cubic capacity/cylinder	3929 cc/12 cylinder
hp/kW	350/256
Construction period	1970 – 1976
Number of produced cars	327

Lamborghini Jarama GTS

As part of the model development the revised Lamborghini Jarama was created in 1972. The car was given the name Jarama 400 GTS, had a higher output and profited from a great number of technical improvements. The Jarama 400 GTS was the last Lamborghini coupé with an engine positioned in front. From 1974, customers also had the possibility to order a three-step automatic gear for the Jarama against a surcharge.

Model	Lamborghini Jamara GTS
Cubic capacity/cylinder	3929 cc/12 cylinder
hp/kW	365/268
Construction period	1973 – 1976
Number of produced cars	150

Lamborghini Urraco

Designed by the studio of famous bodywork stylist Bertone, the new Urraco P 250 with a transversely installed 2.5-litre V8-mid-engine and 220 hp was the centre of attention at the Turin International Motor Show. The P 250 was described as a rare and beautiful combination of balance and harmony, innovation and passion.

Model	Lamborghini Urraco
Cubic capacity/cylinder	2462 cc/8 cylinder
hp/kW	220/161
Construction period	1972 – 1976
Number of produced cars	520

Lamborghini Countach LP 400

Finally, the version of the new Countach LP 400 that was ready to go into mass production was presented to the general public in 1974. The development of this car with the powerful performance had involved a great amount of knowledge gained from racing. The first prototype (the LP 500) of the low riding "racing wedge" had already been presented at the Geneva International Motor Show in 1971. The most spectacular feature of the LP 400 was of course the design of its upward swinging doors.

Model	Lamborghini Countach LP 400
Cubic capacity/cylinder	3929 cc/12 cylinder
hp/kW	375/275
Construction period	1974 – 1978
Number of produced cars	150

Lamborghini Countach LP 500 S

When the Countach had practically reached the stage at which serial production was about to commence, Lamborghini had just mastered the biggest crisis of its company life. The consequences of the oil embargo of 1973 also affected the sports car makers leading to a drastic reduction of sales figures. As a result Ferrruccio Lamborghini was compelled to sell 51 percent of his company.

Model	Lamborghini Countach LP 500 S
Cubic capacity/cylinder	4754 cc/12 cylinder
hp/kW	375/275
Construction period	1982 – 1985
Number of produced cars	320

Lamborghini Countach 25

In 1982, the Countach LP 500 S was launched, a new car with a five-litre engine and 375 hp. In 1985, the racer with the new four-valve technology underwent its third product re-valuation and was renamed LP 500 S QV. When Lamborghini celebrated in 1988 its 25th anniversary as a manufacturer of sports cars, the Countach appeared in its final production version, which was introduced by the manufacturer as a jubilee edition.

Model	Lamborghini Countach 25
Cubic capacity/cylinder	4754 cc/12 cylinder
hp/kW	430/316
Construction period	1988 – 1990
Number of produced cars	657

Lamborghini Silhouette

The Silhouette, built in the construction style of a two-seater coupé with a "targa top", was introduced in 1976 at the International Motor Show in Geneva. It caused quite a sensation especially on account of its dynamically aggressive styling – the square edged wheel cut-out fully corresponded to the prevailing preferences of the car. While the output of the V8 engine was limited to 265 hp, the engine was considered to be absolutely suited for everyday life.

Model	Lamborghini Silhouette
Cubic capacity/cylinder	2996 cc/8 cylinder
hp/kW	265/195
Construction period	1976 – 1980
Number of produced cars	54

Lamborghini Jalpa

In 1981, Lamborghini presented the Jalpa model as a successor of the Silhouette. Its concept consisted once again of a mid-engine sports car whose V8 cylinder engine enabled it to accelerate to 235 km/h. As opposed to comparable competitors (e.g. Ferrari 308 GTB) the Jalpa could not easily hold its own, because a fuel consumption of 17 litres for 100 kilometres was simply too much for a car of this class.

Model	Lamborghini Jalpa
Cubic capacity/cylinder	3485 cc/V8 cylinder
hp/kW	255/187.5
Construction period	1981 – 1988
Number of produced cars	approx. 420

Lamborghini Diablo

In May 1990 the production of the Countach was stopped after 19 years of construction. This freed the way for the Diablo. Production and sales of the Diablo already reached their climax one year later, and the annual financial statements showed that the company operated in the black. However, the crisis of the world market was approaching – in the following year already manufacturers of dream cars had to record a significant drop in sales figures.

Model	Lamborghini Diablo
Cubic capacity/cylinder	5700 cc/12 cylinder
hp/kW	492/362
Construction period	1990 – 1994
Number of produced cars	–

Lamborghini Diablo Jota

The Diablo VT was introduced in 1993. The VT version of the Diablo featured a special clutch that allowed the transmission of power when required also via an all-wheel drive. In autumn of the same year, Lamborghini presented a Diablo Special Edition to celebrate the company's 30th anniversary. Absolute highlight of this edition was the extremely sportily inclined version "Jota".

Model	Lamborghini Diablo Jota
Cubic capacity/cylinder	5700 cc/12 cylinder
hp/kW	580/426
Construction period	1993 – 1995
Number of produced cars	–

Lamborghini Diablo GT

In 1999, Lamborghini presented the world's fastest street sports car of that time, the Diablo GT. The company's experience gained from the development of racing cars was applied to designing this model with a top speed of 338 km/h. As opposed to other Diablo types, this model featured a greatly modified chassis to guarantee optimised driving behaviour even with high speeds.

Model	Lamborghini Diablo GT
Cubic capacity/cylinder	5992 cc/12 cylinder
hp/kW	575/423
Construction period	1999 – 2000
Number of produced cars	80

Lamborghini Diablo GTR

The "Lamborghini Supertrophy" was an in-house race, which was held on many European racetracks. Since 1996, vehicles of the type Diablo SVR – a special racing version – are employed in this race. To meet the requirements of the participants, the company launched the further developed Diablo GTR to set new benchmarks of even more power.

Model	Lamborghini Diablo GTR
Cubic capacity/cylinder	5992 cc/12 cylinder
hp/kW	590/434
Construction period	1999 – 2000
Number of produced cars	30

Lamborghini Diablo SV

Ferruccio Lamborghini was not able to master all crises. In 1972, he withdrew from the company. Since then Automobili Lamborghini S.p.A. changed its owners several times. In 1998, Audi AG took over the exclusive sports car manufacturer. Under their management the Diablo, which had been already produced since 1990, was once more improved and upgraded for the model year 1999.

Model	Lamborghini Diablo SV
Cubic capacity/cylinder	5707 cc/12 cylinder
hp/kW	500/368
Construction period	1999 – 2001
Number of produced cars	–

Lamborghini Diablo Roadster

In 1992, an open variation of the Diablo was presented for the first time to the specialist press on the occasion of the Geneva International Motor Show. Three years later, the debut of the serial version followed in Bologna. This model with four-wheel drive could claim in 1995 to be the world's fastest all-wheel roadster. In 1998, the manufacturer introduced the Roadster SV (Sport Veloce) with conventional propulsion technology as an alternative.

Model	Lamborghini Diablo Roadster
Cubic capacity/cylinder	5707 cc/12 cylinder
hp/kW	530/390
Construction period	1999 – 2001
Number of produced cars	–

Lamborghini Diablo VT

In 1993, a Diablo VT already generated a great amount of media attention when the journalist Leopoldo Canetoli embarked with his VT on a journey to the North Cape. He arrived at his destination, 8533 kilometres away, after only seven days which corresponds to a daily average distance of 1200 kilometres. For the model year 1999, the Diablo VAT , which had already been built since 1991, was submitted once again to a thorough model upgrade.

Model	Lamborghini Diablo VT
Cubic capacity/cylinder	5707 cc/12 cylinder
hp/kW	492/362
Construction period	1991 – 1999
Number of produced cars	–

Lamborghini Diablo 6.0

After a total of eleven years of production, Lamborghini bid the Diablo class farewell in 2001. Since 1990, about 3000 units had left the factory halls in Sant Agata. The good-bye present from Lamborghini to its customers was called Diablo 6.0 Special. In line with the high-tech style of the car, the interior of this model featured many details made of carbon.

Model	Lamborghini Diablo 6.0
Cubic capacity/cylinder	5707 cc/12 cylinder
hp/kW	575/423
Construction period	2001
Number of produced cars	45

Lancia Stratos

Often interesting racing cars originate from standard vehicles that have been tuned – however, the development of the Stratos went the opposite way. In this case, a racing sports car was tamed to make it street-legal. In the street version, the speedometer needle only came to a halt at 230 km/h. To save weight, the bodywork of the Stratos was made of synthetic material.

Model	Lancia Stratos
Cubic capacity/cylinder	2418 cc/6 cylinder
hp/kW	190/140
Construction period	1973 – 1976
Number of produced cars	502

Lancia Beta Spider

In 1973, Lancia introduced the Beta model. It was offered in the body styles coupé, sports station wagon (HPE) and Spider. All three were interesting cars, especially the Spider. However, anyone who expected this to be a completely open car, was badly disappointed, because with this model only the middle of the roof section was removable (similar to the Targa concept).

Model	Lancia Beta Spider
Cubic capacity/cylinder	1992 cc/4 cylinder
hp/kW	135/99
Construction period	1973 – 1985
Number of produced cars	–

Lancia Beta Montecarlo

In spite of its racing successes, the economic trouble of Lancia could no longer be concealed in the late 1960s – the company was taken over by Fiat. Under the new management first the Lancia Beta, a four door saloon was launched. Sports car fans could fall back on the Beta Coupe, the Beta Spider or later the Beta Montecarlo.

Model	Lancia Beta Montecarlo
Cubic capacity/cylinder	1995 cc/4 cylinder
hp/kW	120/88
Construction period	1975 – 1982
Number of produced cars	–

Lancia Thema 8.32

With the "Thema" saloon introduced in 1984, Lancia was able to achieve a gigantic commercial success. The big four-door model could even be driven in a sportive manner if one drove the later released version 8.32. In this version, a Ferrari V8 engine was the main feature. In line with the Ferrari look, the 8.32 was equipped with a very noble interior.

Model	Lancia Thema 8.32
Cubic capacity/cylinder	2927 cc/8 cylinder
hp/kW	205/151
Construction period	1986 – 1991
Number of produced cars	–

Lancia Delta HF Integrale 16 V

Undoubtedly, the Fulvia coupé established in the late 1960s Lancia's great rally tradition. With Lancia Stratos and the legendary Type 037, Lancia was even able to gain the first world title for Turin, and this was still far away from the end of the rope. Later on, the all-wheel-drive models Delta HF 4 WD and HF 4 Integrale won the manufacturer's trophy another six times!

Model	Lancia Delta HF Integrale 16V
Cubic capacity/cylinder	1995 cc/4 cylinder
hp/kW	210/155
Construction period	1989 – 1991
Number of produced cars	–

Maserati Bora

The US market has always been a lucrative field of activity for European sports car manufacturers. Therefore, exotics like the Maserati Bora (maximum speed 270 km/h) always had to comply with the strict exhaust regulations there. At the beginning of the 1970s the laws were drastically tightened once again. Incidentally, today Bora coupés fetch particularly high prices due to their small production volumes.

Model	Maserati Bora
Cubic capacity/cylinder	4719 cc/8 cylinder
hp/kW	310/227
Construction period	1971 – 1980
Number of produced cars	571

Maserati Khamsin

The Khamsin, named after an Egyptian desert wind, incorporated a concept that was almost identical to the Indy. However, this two-seater with a maximum speed of 270 km/h and a hatchback could be equipped with a rear jump seat bank. Citroen contributed, incidentally, the high-pressure hydraulic systems for the steering, the brakes and the clutch.

Model	Maserati Khamsin
Cubic capacity/cylinder	4930 cc/8 cylinder
hp/kW	320/235
Construction period	1972 – 1982
Number of produced cars	421

Maserati Merak

In 1968, Maserati and Citroen embarked on a joint collaboration whose contributions to the automotive world included the Maserati Merak. This model incorporated many Citroen parts. In addition to the three-litre Merak, Maserati also produced a two-litre variation for the Italian market. Because both models were available at an affordable price, they always provided full order books.

Model	Maserati Merak
Cubic capacity/cylinder	2965 cc/6 cylinder
hp/kW	220/161
Construction period	1972 – 1983
Number of produced cars	approx. 1800

Maserati Kyalami

In the Zulu language "Kyalami" means "my home". This model name was a reference to the first race on the Kyalami race track in South Africa. Optically very strongly resembling the De Tomaso Longchamp, the Kyalami had a wheel base of 260 centimetres, while its wheel suspensions were identical to those of the Maserati Khamsin.

Model	Maserati Kyalami
Cubic capacity/cylinder	4930 cc/8 cylinder
hp/kW	280/206
Construction period	1978 – 1983
Number of produced cars	150

Maserati Spyder

In 1984, three years after the introduction of the Biturbo series, Maserati presented the Spyder to the great joy of all convertible drivers. In comparison to the coupe, the Spyder (maximum speed 210 km/h) received a wheel base that was shortened by about twelve centimetres. The propulsion technology was taken over without any changes from the Biturbo – a two-litre engine for the domestic market, 2.5 litres for the export model.

Model	Maserati Spyder
Cubic capacity/cylinder	2491 cc/6 cylinder
hp/kW	192/141
Construction period	1984 – 1988
Number of produced cars	1049

Maserati 3200 GT

Giorgetto Giugiaro had already designed a great number of Maserati models in the past. The shape of the 3200 GT four-seater fastback coupé (266 centimetres of wheel base) likewise originated on his drawing board. As part of the model development, the "Assetto Corsa" variant with a lower chassis was introduced two years later.

Model	Maserati 3200 GT
Cubic capacity/cylinder	3217 cc/8 cylinder
hp/kW	326/240
Construction period	1988 – 2001
Number of produced cars	–

Maserati Shamal

The Shamal, named after a Mesopotamian desert wind, was designed by Marcello Gandini who had already become famous for designing the Maserati Khamsin. The Shamal, a so called 2 + 2-seater, was equipped with a V8 bi-turbo engine. In addition, it was also the first Maserati model equipped with a six-gear transmission.

Model	Maserati Shamal
Cubic capacity/cylinder	3217 cc/8 cylinder
hp/kW	326/240
Construction period	1990 – 1996
Number of produced cars	–

Maserati 222 4v

In comparison to the Maserati 222 R, which had 50 hp less power, the 222 4v featured a more powerful athletic engine with four-valve technology and four overhead camshafts underneath its bonnet. Measuring only 420 centimetres in length, the 222 4v moved on 16 inch wheels at a top speed of 255 km/h. Maserati called the body style of this model, incidentally, a "notchback coupé".

Model	Maserati 222 4v
Cubic capacity/cylinder	2790 cc/6 cylinder
hp/kW	278/204
Construction period	1991 – 1994
Number of produced cars	130

Saab Sonett III

With the Sonnet III, launched in 1970, Saab ventured one last attempt to position this model in the market. Compared to the Sonett II, the new model was updated technically and optically and was now equipped with a four-cylinder engine, a modern floor-mounted gear change, as well as light alloy rims. The expenditure had been worthwhile – finally, the Sonett became interesting to Saab customers.

Model	Saab Sonett III
Cubic capacity/cylinder	1498 cc/4 cylinder
hp/kW	65/48
Construction period	1970 – 1974
Number of produced cars	8336

Saab 900 Cabriolet

The positive resonance of the general public to an open concept car presented by Saab in 1983 encouraged the company to implement the project. Three years later they were ready: the first slightly sporty Saab 900 Cabriolet was sold to an American customer. At the same time this was the beginning of a success story spanning two whole decades.

Model	Saab 900 Cabriolet
Cubic capacity/cylinder	1985 cc/4 cylinder
hp/kW	160/118
Construction period	1986 – 1994
Number of produced cars	49.000 (all versions)

Saab 900 Cabriolet

Saab had hesitated for a long time before taking the decision to invest production capacities into a convertible. Therefore, initially it only planned the production of 400 units exclusively for the US market. However, due to the overwhelming demand the Saab 900 Cabriolet was soon sold around the globe. The second generation that came off the conveyor belt as of 1994 also provided Saab with full order books.

Model	Saab 900 Cabriolet
Cubic capacity/cylinder	1985 cc/4 cylinder
hp/kW	160/118
Construction period	1986 – 1994
Number of produced cars	49.000 (all versions)

Saab 900 Turbo Cabriolet

The second generation of the 900 convertible series was introduced to the market in 1994. According to the slogan – "We have not invented the convertible, but we have turned it into a Saab!" the new model corresponded precisely to the wishes of its individualist customers. Finally, the folding roof disappeared completely under a snug-fitting lid in the boot.

Model	Saab 900 Turbo Cabriolet
Cubic capacity/cylinder	1988 cc/4 cylinder
hp/kW	175/129
Construction period	1994 – 2003
Number of produced cars	140.500 (all versions)

Saab 900 SE

In some European markets, for instance in Great Britain, Sweden and the Netherlands, the convertible Saab model repeatedly generated more than 50 percent of sales. In Germany, the Saab convertible at times even accounted for half of all Saab sales. Models of the first generation, incidentally, have already recently become collector's items in the future classic car scene.

Model	Saab 900 SE
Cubic capacity/cylinder	1998 cc/4 cylinder
hp/kW	150/110
Construction period	1994 – 2003
Number of produced cars	140.500 (all versions)

Felber FF

The Swiss native Willy Felber created a number of sports cars in the 1970s that certainly were not to everybody's linking. His Felber FF whose body shape resembles the early Ferrari 125 a little, was based, for instance, on a Ferrari chassis. The engine also formerly served Ferrari cars. Now it provided the FF with enough power to let it achieve a top speed of at least 225 km/h.

Model	Felber FF
Cubic capacity/cylinder	3967 cc/12 cylinder
hp/kW	300/220
Construction period	1974 – 1979
Number of produced cars	approx. 12

Monteverdi Shark 450 GTS

As a further development of the Hai 450 SS unique piece shown in 1970, three years later Peter Monteverdi introduced the Hai 450 GTS. Like his predecessor, this version was also conceived as a mid-engine-sports coupé, however, the GTS was equipped with a more comfortable interior. Incidentally the production figures of the exclusive Monteverdi cars were never announced.

Model	Monteverdi Shark 450 GTS
Cubic capacity/cylinder	6974 cc/8 cylinder
hp/kW	390/287
Construction period	1973
Number of produced cars	–

Rinspeed Roadster R

In 1977, Frank M. Rinderknecht founded the Rinspeed Garage. He first of all established himself as a tuning and modifying expert before introducing his own brand in 1995 with the "Roadster R". The appearance of this exclusive car was reminiscent of the car shapes of the 1960s combined with modern elements. Its bold look was aimed at a buyer segment that wanted to enjoy life and fresh air.

Model	Rinspeed Roadster R
Cubic capacity/cylinder	4995 cc/8 cylinder
hp/kW	218/160
Construction period	1995
Number of produced cars	–

Rinspeed Yello Talbo

The Yello Talbo was a symbiosis of nostalgic and futuristic elements. Its unusual appearance dates back to the Talbot Lago 150 SS of 1938. The interior of this luxury car was finished with noble materials by "En Soie". Thanks to a charged V8 engine, the car could reach the 100 km/h mark in only 5.5 seconds.

Model	Rinspeed Yello Talbo
Cubic capacity/cylinder	4995 cc/8 cylinder
hp/kW	320/235
Construction period	1996
Number of produced cars	–

Rinspeed Mono Ego

The Rinspeed Mono Ego may boast to have been the world's first single-seater car that has been awarded a street registration. The exclusive model was launched on the occasion of the 20th company anniversary of Rinspeed AG. For Frank M. Rinderknecht this car represented a symbiosis of traditional beauty and future-oriented technology.

Model	Rinspeed Mono Ego
Cubic capacity/cylinder	4995 cc/8 cylinder
hp/kW	410/301
Construction period	1997
Number of produced cars	–

Rinspeed E-Go Rocket

The Rinspeed E-Go Rocket, an exclusive, fully street-legal vehicle, had an appearance reminiscent of the legendary salt flats race record vehicles of the 1940s. This automotive piece of art shown for the first time in 1998 at the Geneva Salon (200 km/h maximum speed) was only made by order, and as a special feature customised baggage was delivered with the car.

Model	Rinspeed E-Go Rocket
Cubic capacity/cylinder	4995 cc/8 cylinder
hp/kW	410/301
Construction period	1998
Number of produced cars	–

Cadillac Evoq

The Evoq Roadster was intentionally designed low-key and according to the motto of "less is more". This meant that, among other innovations, door handles did not exist anymore since the car had central locking without keys. The technical highlights of the concept car included a three-part hard top that could completely disappear at the touch of a button.

Model	Cadillac Evoq
Cubic capacity/cylinder	4200 cc/8 cylinder
hp/kW	410/302
Construction period	1999
Number of produced cars	unique piece

Chevrolet Corvette Stingray

After the model change in 1974, the Corvette said goodbye to overly abundant power output – the "big block engine" with 7.4 litres of cubic capacity was cancelled. Its place was taken by a 5.7-litre V8 engine, which surprisingly did not adversely affect the sales figures, because with this engine the car could also easily reach a top speed of 200 km/h.

Model	Chevrolet Corvette Stingray
Cubic capacity/cylinder	5733 cc/8 cylinder
hp/kW	220/162
Construction period	1974 – 1982
Number of produced cars	–

Chevrolet Corvette

The proportions were right: with a height of 119 centimetres and a total length of 448 centimetres, it looked as if the 180 centimetre wide Corvette was tightly huddled against the street. Quite deliberately the fourth generation of this model from 1984 to 1996 was clad in a simple and understated body style, its design corresponded precisely to the preferences of the time.

Model	Chevrolet Corvette
Cubic capacity/cylinder	5710 cc/8 cylinder
hp/kW	205/151
Construction period	1984 – 1996
Number of produced cars	–

Chevrolet Corvette

Under the bonnet of the Corvette reigned a beefy V8 cylinder engine, the so-called "small block." Upon closer examination it could be seen that many of the parts attached to the engine were manufactured in a very modern fashion. The casing of the brake force booster consisted of synthetic material, while the mounting of the dynamo as well as parts of the air-conditioning compressor were made of aluminium.

Model	Chevrolet Corvette
Cubic capacity/cylinder	5710 cc/8 cylinder
hp/kW	205/151
Construction period	1984 – 1996
Number of produced cars	–

Chevrolet Corvette

Even if it was not immediately apparent due to its low waistline – the fourth Corvette generation was actually a little smaller than its curvaceous predecessor. In addition to the coupé and following a long break, as of 1986 Chevrolet built the Corvette also as a convertible once again. Apparently a good idea because nearly one in three buyers opted for this slightly more expensive alternative.

Model	Chevrolet Corvette
Cubic capacity/cylinder	5710 cc/8 cylinder
hp/kW	330/243
Construction period	1986 – 1996
Number of produced cars	–

Chevrolet Corvette

In 1997, the US carmaker introduced a new highlight of the Corvette legendary sports car series to the general public. Now in its fifth generation, the sports car appeared with an especially aerodynamic synthetic body. With the efforts of chief engineer Dave Hill a cd value of 0.29 was reached. Another innovation was the even more stable steel frame underbody.

Model	Chevrolet Corvette
Cubic capacity/cylinder	5700 cc/8 cylinder
hp/kW	344/253
Construction period	1997 – 2004
Number of produced cars	–

Chevrolet Corvette

The many improvements of the fifth Corvette generation included the enlargement of the boot trunk, which became almost twice as large as that of the predecessor's model. Another advantage: the seat position could be saved in three settings with the help of a memory function, so that everyone who used the car could call up his/her individual position by the simple press of a button.

Model	Chevrolet Corvette
Cubic capacity/cylinder	5700 cc/8 cylinder
hp/kW	344/253
Construction period	1997 – 2004
Number of produced cars	–

Chevrolet Corvette

In the year 1999, for the first time in the model's history, Corvette buyers were given the choice between three body variations. In addition to the coupé with a removable roof middle section, and the convertible, a hard top-version was also available from the retailers. Unfortunately, this interesting alternative was exclusively available in the US market.

Model	Chevrolet Corvette
Cubic capacity/cylinder	5700 cc/8 cylinder
hp/kW	344/253
Construction period	1997 – 2004
Number of produced cars	–

Chevrolet Camaro

The Camaro, which was introduced in 1967, was Chevrolet's answer to the Ford Mustang. According to US standards, the Camaro was a relatively small car. Only in the course of model development did it become a powerful car with good acceleration values. The second Camaro generation, produced as of 1970, barely had any sales prospects in the European market due to the pending oil crisis.

Model	Chevrolet Camaro
Cubic capacity/cylinder	4094 cc/6 cylinder
hp/kW	100/73
Construction period	1970 – 1975
Number of produced cars	–

Clénet Series II

In the 1970s and 1980s, the US "in vogue scene" enjoyed being seen in so-called neoclassic vehicles. Clénet was among the manufacturers who created such cars. Californian native Alain Clénet founded the company in 1975. Until the demise of the company in 1983, Clénet employed a staff of up to 130 individuals.

Model	Clénet Series II
Cubic capacity/cylinder	5770 cc/8 cylinder
hp/kW	142/104
Construction period	1979 – 1983
Number of produced cars	–

Dodge Charger Super Bee

Dodge supplied the Super-Bee models with quite a special mascot: on the radiator grille and on the rear mudguards there was an emblem featuring the "Super Bee" which referred to the V8 engine at work under the bonnet. However in the basic version it consisted of only a 4.5-litre engine, but 7.2 litres of cubic capacity were also optionally available.

Model	Dodge Charger Super Bee
Cubic capacity/cylinder	7195 cc/8 cylinder
hp/kW	390/287
Construction period	1970 – 1971
Number of produced cars	–

Dodge Charger SE

The power engines of the early 1970s started to disappear one by one as a result of the increasingly stringent exhaust laws in the USA. The largest remaining Charger engines included the 7.2-litre V8 with 185 hp or the 5.9-litre version with 190 hp. In addition, the Charger model was now only offered as a two-door hard top-version to compete with Chevrolet's "Monte Carlo".

Model	Dodge Charger SE
Cubic capacity/cylinder	5898 cc/8 cylinder
hp/kW	190/140
Construction period	1976 – 1977
Number of produced cars	–

Dodge Daytona

Based on a modified K-Car chassis, Dodge introduced the Daytona Sports Coupé in the model year 1984. The interesting thing about this car was the fact that the Daytona was the USA's first front-wheel drive turbo car! However, whether as a turbo or as a standard design version, both turned out to be rather slow burners in terms of [illegible] figures.

Model	Dodge Daytona
Cubic capacity/cylinder	2199 cc/4 cylinder
hp/kW	142/104
Construction period	1984 – 1993
Number of produced cars	–

Dodge Viper RT/10 Cabriolet

When Dodge presented in 1989 a sports concept car entitled "Viper" at various auto shows, the general public reacted so enthusiastically that Dodge Division was left with no recourse but to serially produce this intriguing two-seater. In the summer of 1992 the first units were delivered – exclusively painted in red.

Model	Dodge Viper RT/10 Cabriolet
Cubic capacity/cylinder	7990 cc/10 cylinder
hp/kW	364/268
Construction period	1992 – 1996
Number of produced cars	–

Dodge Viper RT/10 Cabriolet

For 1992, Dodge had planned to produce only 200 units of the viper – so of course this small series was sold off immediately. Everybody wanted to have this Roadster, which was strong as an ox with a V10 engine, some customers were even ready to acquire the viper above its list price. Nevertheless, it took some time until Dodge decided to increase the annual production to about 1400 to 2000 units.

Model	Dodge Viper RT/10 Cabriolet
Cubic capacity/cylinder	7990 cc/10 cylinder
hp/kW	364/268
Construction period	1992 – 1996
Number of produced cars	–

Dodge Viper GTS Coupé

Already in 1996 a coupé variation, the GTS, was juxtaposed to the roadster. In terms of engine technology, nothing had changed compared to the roadster. Yet the GTS was quite a different car – it weighed far less as a result of many construction improvements that definitely benefited its agility.

Model	Dodge Viper GTS Coupé
Cubic capacity/cylinder	7990 cc/10 cylinder
hp/kW	383/282
Construction period	1996 – 2003
Number of produced cars	–

Ford Thunderbird Concept Car

Shortly after the 1997 announcement that the production of the Thunderbird would be ceased – the wheels of production started moving within the Ford Motor Company. A team of technicians developed a new Thunderbird concept, which received great attention two years later in the form of a design concept car.

Model	Ford Thunderbird Concept Car
Cubic capacity/cylinder	3900 cc/8 cylinder
hp/kW	243/179
Construction period	1999
Number of produced cars	design study

Dodge Viper GTS-R

In 1996, with the introduction of the GTS coupé, the GTS-R racing version also became a reality. A prototype of this variation (shown here) was already introduced in 1989. These racing models with up to 750 hp scored various victories, also in Le Mans. Due to this great success, Dodge introduced a scaled down coupé special series for private drivers limited to 100 units.

Model	Dodge Viper GTS-R
Cubic capacity/cylinder	7990 cc / 10 cylinder
hp/kW	750/551
Construction period	1996
Number of produced cars	–

Phoenix Cobra

When the last "real" Cobras were sold at Ford dealers in 1973, it did not take long before profiteers copied this sports car in a more or less successful fashion. One of the models, which one may classify among the successful replicas, was the Phoenix Cobra. It was appropriately equipped with Ford's "Big-Block" engine and received a roll bar as well as chrome plated "sidepipes".

Model	Phoenix Cobra
Cubic capacity/cylinder	7485 cc/8 cylinder
hp/kW	389/286
Construction period	1981 – 1984
Number of produced cars	–

Plymouth Prowler

In 1993, Chrysler presented the Plymouth Prowler as a concept car. In the middle of 1997 the mass production started for the American market where it immediately revived not only the hot-rod but also the sports cars scene. The 419 centimetre-long Roadster did not only feature a fascinating courageous design, but also a great amount of futuristic technology.

Model	Plymouth Prowler
Cubic capacity/cylinder	3518 cc/6 cylinder
hp/kW	214/157
Construction period	1997 – 2002
Number of produced cars	–

Plymouth Prowler

Actually, the Prowler was intended to remain a concept study. However, based on the positive feedback, Chrysler was persuaded to build the car in series. The Prowler body was made of 98 percent aluminium, while the remainder consists of carbon reinforcement. Both materials were connected to each other by special welding and adhesive techniques.

Model	Plymouth Prowler
Cubic capacity/cylinder	3518 cc/6 cylinder
hp/kW	214/157
Construction period	1997 – 2002
Number of produced cars	–

Plymouth Prowler

The basic equipment of the 210 km/h-fast Prowler included its 20-inch sized and wide back rear tyres, as well as the air-conditioning, the passenger seat air bag and a cassette radio. In May 2002, the last car of the series came off the conveyor belt and was auctioned for charitable purposes. It generated a total amount of USD 175.000.

Model	Plymouth Prowler
Cubic capacity/cylinder	3518 cc/6 cylinder
hp/kW	214/157
Construction period	1997 – 2002
Number of produced cars	–

Vector W2

The industrial designer Gerald Wieger was among the few people in the world who were able to make the dream of making his own car come true. As of 1980, he planned to annually produce in his California plant about 300 units of his extremely low riding 375 km/h-fast W2 model (based on Corvette technology and equipped with two turbochargers) – an extremely unrealistic value.

Model	Vector W2
Cubic capacity/cylinder	5733 cc/8 cylinder
hp/kW	600/441
Construction period	1980 – 2000
Number of produced cars	–

Holden Torona GTR-X

In 1967, Holden introduced the Torona, an expandable medium-sized car model that was already complemented three years later by a sportily inclined model. As opposed to the smallest saloon (1.1 litres cubic capacity and four-cylinder engine) the sports variation (Torona GTR-X) profited from a six-cylinder-engine of the three-litre-class.

Model	Holden Torona GTR-X
Cubic capacity/cylinder	3043 cc/6 cylinder
hp/kW	162/119
Construction period	1970–1972
Number of produced cars	–

Holden Torona A9X

The Holden A9X, Australia's first and at that time only high capacity sports cars debuted in the summer of the year 1977. In terms of appearance, the A9X was a little reminiscent of the Opel Kadett, which was not surprising, since after all, both brands belonged to the GM Group. The A9X was serially delivered with a bonnet painted in black on which a backward facing air scoop sat enthroned.

Model	Holden Torona A9X
Cubic capacity/cylinder	[illegible] cc/[illegible] cylinder
hp/kW	219/161
Construction period	1977–1978
Number of produced cars	

Honda CR-X

Many car enthusiasts who were in search of an alternative to the "GTI" or "GTE" in the late 1980s, found what they were looking for in the CR-X of the second generation offered by Honda. This cult car of which driving licence novices dreamed and that could be excellently tuned was above all else fast. The 130-hp version reached a maximum speed of 205 km/h, while the 150-hp engine even achieved a maximum speed of 222 km/h.

Model	Honda CR-X
Cubic capacity/cylinder	1590 cc/4 cylinder
hp/kW	150/110
Construction period	1987 – 1992
Number of produced cars	–

Honda CR-X 1.6i

Based on the fifth generation of the Honda Civic, the probably last CR-X model for the time being was introduced in 1992. As opposed to its predecessors, this variation consisted of a strictly two-seater with a removable targa roof. As a result of its extensive popularity, the CR-X (as well as the discontinued NSX) continued to enjoy great popularity even as a used car.

Model	Honda CR-X 1.6i
Cubic capacity/cylinder	1590 cc/4 cylinder
hp/kW	125/92
Construction period	1992–1998
Number of produced cars	[illegible]

Honda CR-X 1.6 VTi

Honda introduced the CR-X with two engine levels – as an alternative to the 1.6i version with 125 hp they also offered the stronger and substantially more agile 1.6 VTi (VTEC engine) with 160 hp. In addition, Honda also launched some special models, for example, the especially exclusively equipped "Motegi" that was only built 1200 times.

Model	Honda CR-X 1.6 VTi
Cubic capacity/cylinder	1595 cc/4 cylinder
hp/kW	160/118
Construction period	1992 – 1998
Number of produced cars	17.400

Honda NSX

When Honda launched its NSX series in 1988, the development engineers had only one purpose in mind – the NSX should become a modern sports car whose road performance level was fully equivalent to the quality of traditional sports cars. Therefore, the only possible body style was a low silhouette with far-stretched passenger's cabin.

Model	Honda NSX
Cubic capacity/cylinder	2977 cc/6 cylinder
hp/kW	274/201
Construction period	1990 – 1996
Number of produced cars	–

Honda NSX

During the long production period of the NSX there never was a reason to carry out larger cosmetic changes as part of model development. The design was correct right from the start. Another reason why the sports car was continuously produced without any changes throughout the years – it was assembled to a great extent by manual labour in a "special factory" set-up.

Model	Honda NSX
Cubic capacity/cylinder	2977 cc/6 cylinder
hp/kW	274/201
Construction period	1990 – 1996
Number of produced cars	–

Honda NSX

From the start, Honda's mid-engine sports car was considered to be a reliable car that could be used comfortably in everyday situations. In every model level, its V6 engine disposed of amply sufficient torque, which made the NSX an agile high-tech car. A continued performance beyond the 200,000-kilometre mark (with the first engine) was not unusual.

Model	Honda NSX
Cubic capacity/cylinder	3179 cc/6 cylinder
hp/kW	280/206
Construction period	1997 – 2001
Number of produced cars	–

Honda NSX-T

Although Honda always remained loyal to the NSX basic concept and rejected larger rejuvenating measures to its appearance, the technology of the NSX was always updated "under the bonnet". The model range was extended only once, by the NSX T. This model consisted of a targa version with a removable middle roof section which was introduced alongside the coupé.

Model	Honda NSX-T
Cubic capacity/cylinder	3179 cc/6 cylinder
hp/kW	280/206
Construction period	1997 – 2001
Number of produced cars	–

Honda Integra Type R

The Integra Type R that was actually only intended for the US and Japanese markets was also sold between 1998 and 2001 in Germany. The model history of this series (originally a middle class saloon) goes back to 1985. With the R-version Honda released a variation with a sports chassis, spoiler kit, Recaro seats and numerous other sporty details.

Model	Honda Integra Type R
Cubic capacity/cylinder	1795 cc/4 cylinder
hp/kW	190/140
Construction period	1998 – 2001
Number of produced cars	–

Honda S 2000

Honda presented the Roadster S 2000 on the occasion of the company's 50th anniversary. Shortly after its debut, the pretty car was the first car by a Japanese manufacturer to be honoured with the title "convertible of the year". Under the bonnet of the dynamic two-seater a two-litre four-cylinder engine with variable valve technology (VTEC) provided quite some bite – and a maximum speed of 240 km/h.

Model	Honda S 2000
Cubic capacity/cylinder	1997 cc/4 cylinder
hp/kW	241/177
Construction period	from 1999
Number of produced cars	–

Honda S 2000

It could not be stated any better: "The S 2000 provides genuine driving pleasure and additionally is totally in line with our business philosophy which sets unequivocal priorities in the areas of security and environmental protection", declared Honda president Hiroyuki Yoshino on the occasion of the official formal launch of the serial production in the summer of 1999.

Model	Honda S 2000
Cubic capacity/cylinder	1997 cc/4 cylinder
hp/kW	241/177
Construction period	from 1999
Number of produced cars	–

Mazda RX-3

In 1967, Mazda began to export its cars to Europe, and only five years later the group founded "Mazda Motors Germany ". Initially, five models were imported, including the Mazda-RX-3 Coupé equipped with a planetary piston engine. Mazda classified, incidentally, all cars with planetary piston engines in the category of sports cars.

Model	Mazda RX-3
Cubic capacity/cylinder	2 x 491 cc
hp/kW	105/77
Construction period	1972 – 1974
Number of produced cars	–

Mazda RX-5

Japanese sports cars have a much longer tradition than was generally believed. Mazda could only establish itself in this class with the RX-7. However, its predecessors models were also among the most spirited cars that could be purchased in the 1970s. The RX-5 equipped with a Wankel engine was sold in Japan, incidentally, under the name Cosmo.

Model	Mazda RX-5
Cubic capacity/cylinder	chamber volumes 2292 cc
hp/kW	125/92
Construction period	1975 – 1979
Number of produced cars	–

Mazda RX-7

With the RX-7 presented in spring 1978, Mazda introduced a new sports car generation whose concept remained constant as part of the model development until the year 2002. Mazda equipped the "RX 7" coupé, also known as "Savanna", with a planetary piston engine – this concept developed by Felix Wankel was improved on by Mazda.

Model	Mazda RX-7
Cubic capacity/cylinder	chamber volumes 2292 cc
hp/kW	105/77
Construction period	1978 – 1982
Number of produced cars	–

Mazda RX-7 GS

The size of the 2+2-seater sports coupé RX-7 corresponded roughly to that of the Porsche 924. Just like the Porsche, the Mazda featured modern sleeping eyes headlights and was serially equipped with more generous basic equipment. The GS version, which could be recognised by lateral lining strips, was even equipped a little more luxuriously (five speed gearbox, interval windshield wiper, sporty centre console, etc.).

Model	Mazda RX-7 GS
Cubic capacity/cylinder	chamber volumes 2292 cc
hp/kW	105/77
Construction period	1978 – 1982
Number of produced cars	–

Mazda RX-7

With an improved chassis and updated bodywork the third production series of the RX-7 was launched in 1985. In addition to the coupé, Mazda also built a convertible version as of 1989. In general, the engine produced an output of 150 hp. The alternatively offered turbo versions (180 or 200 hp) enabled the RX-7 to achieve a top speed of 240 km/h.

Model	Mazda RX-7
Cubic capacity/cylinder	chamber volumes 2292 cc
hp/kW	150/110
Construction period	1985 – 1991
Number of produced cars	–

Mazda RX-7

Following a skilfully carried out face-lift, the first RX 7 models of the second generation came off the conveyor belt in late 1981. Because of the increased performance and the use of bigger wheels, the car with a maximum speed of more than 200 km/h also received disc brakes in the rear – Mazda's calculated effort of producing a worthy opponent to the Porsche 924 seemed to pay off.

Model	Mazda RX-7
Cubic capacity/cylinder	chamber volumes 2292 cc
hp/kW	115/85
Construction period	1981 – 1985
Number of produced cars	–

Mazda RX-7

Although the last production series of the RX-7 was manufactured from 1991 to 2002, the elegant sports coupé was already discontinued in some of its markets in the mid-1990s. This was due to the austere exhaust regulations in those countries. In Germany, where the RX-7 cost about EUR 40.000 at that time, the sales were discontinued in 1996.

Model	Mazda RX-7
Cubic capacity/cylinder	chamber volumes 2292 cc
hp/kW	200/147
Construction period	1991 – 2002
Number of produced cars	–

Mazda Eunos Cosmo S

In the European market, the Mazda Eunos Cosmo belonged to the unknown Greats, because the car that was in general only equipped with a right-hand drive system was manufactured exclusively for the Japanese and Australian markets. Mazda offered the model that was equipped with a Wankel engine and two turbochargers in two output levels (Type S or Type Sport E).

Model	Mazda Eunos Cosmo S
Engine	2 rotor Wankel engine
Cubic capacity/cylinder	chamber volumes 1302 cc
hp/kW	230/169
Construction period	1990 – 1995
Number of produced cars	approx. 3800

Mazda Eunos Cosmo Sport E

For the Eunos that was conceived as a sporty two-seater only a few extras were available for a surcharge. Everything that made driving more comfortable and more pleasant (four-gear automatic transmission, leather seats, radio, etc.), already belonged to the car's basic equipment. Incidentally, the maximum speed of the Eunos was electronically limited to 185 km/h. Without the limiter, the car could achieve a maximum speed of about 250 km/h.

Model	Mazda Eunos Cosmo Sport E
Engine	3 rotor Wankel engine
Cubic capacity/cylinder	chamber volumes 1962 cc
hp/kW	300/221
Construction period	1990 – 1995
Number of produced cars	approx. 4000

Mitsubishi 3000 GT

The Mitsubishi 3000 GT, which was constructed identically to the American Dodge Stealth, was sold in some countries under the model name GTO. The cars manufactured from 1990 to 1994 received pop-up headlights. In order to comply with all registration regulations, Mitsubishi integrated the parking light for the version offered in Europe into the front bumper.

Model	Mitsubishi 3000 GT
Cubic capacity/cylinder	3000 cc/6 cylinder
hp/kW	286/210
Construction period	1990 – 1994
Number of produced cars	–

Mitsubishi 3000 GT

Thanks to a spirited double turbo engine, the Mitsubishi 3000 GT reached a maximum speed of 250 km/h. This made the model at the time of its market launch the fastest and strongest serial vehicle that had ever come off the conveyor belt of Japanese manufacturers. The version made as of 1995 can be easily recognised, incidentally, by its modern ellipsoid headlights.

Model	Mitsubishi 3000 GT
Cubic capacity/cylinder	3000 cc/6 cylinder
hp/kW	286/210
Construction period	1995 – 2000
Number of produced cars	–

Nissan Datsun 260 Z

Between 1972 and 1973, to comply with stricter US exhaust regulations, Nissan already increased the cubic capacity from 2.4 litres to 2.6 litres and subsequently lowered the power output to 126 hp. The car, which was called 260 Z as of 1975, continued to be offered as an elegant two-seater or a so-called 2 + 2 seater with a larger bodywork that was extended by 30 centimetres.

Model	Nissan Datsun 260 Z
Cubic capacity/cylinder	2547 cc/6 cylinder
hp/kW	126/93
Construction period	1975 – 1979
Number of produced cars	–

Nissan Datsun 280 ZX

For the model year 1975, Nissan increased the cubic capacity to 2.8 litres. The output of the US version was increased, despite the lowering of the compression for reasons of exhaust technology – the introduction of a "Bosch-L-Jetronic" instead of the previous carburettor contributed significantly to this increase. The 280 Z versions for the California market were already equipped with a catalytic converter!

Model	Nissan Datsun 280 ZX
Cubic capacity/cylinder	2753 cc/6 cylinder
hp/kW	140/103
Construction period	1979 – 1984
Number of produced cars	

Nissan Datsun 260 Z

Following the successful market launch of the Z class, soon the production at the Hiratsuka Nissan plant operated in three shifts around the clock. Occasionally, interested buyers had to even accept delivery times of up to twelve months. In competitive racing, the "Z" had also been able to distinguish itself in the meantime – in 1973 it won the East African Safari Rallye in Kenya.

Model	Nissan Datsun 260 Z
Cubic capacity/cylinder	2547 cc/6 cylinder
hp/kW	126/93
Construction period	1975 – 1979
Number of produced cars	–

Nissan Datsun 280 ZX T

Starting in 1978, a slightly revised 280 ZX came off the conveyor belts. As before, it was offered as a two-seater as well as a 2+2-seater. At first glance, the new model looked identical to its predecessor, because a single metal part was taken over from the old model. The new 280 ZX was the first Nissan that was subjected to the wind tunnel to achieve good aerodynamics – its drag coefficient was c_d = 0.38.

Model	Nissan Datsun 280 ZX T
Cubic capacity/cylinder	2753 cc/6 cylinder
hp/kW	200/147
Construction period	1979 – 1984
Number of produced cars	–

Nissan 300 ZX

The discontinuation of the straight six-cylinder engine and the adopting of six cylinders in a V arrangement resulted in 1983 in the third Nissan generation called 300 ZX. The shorter construction style coupled with an enlarged engine with three litres of cubic capacity enabled the designers to pull the front part of the wedge-shaped car very far down. As a pleasant result, the c_d value dropped to 0.31.

Model	Nissan 300 ZX
Cubic capacity/cylinder	2960 cc/6 cylinder
hp/kW	170/125
Construction period	1984 – 1988
Number of produced cars	–

Nissan 300 ZX T

In summer 1985, Nissan followed with a turbo-version of the 300 ZX. This model was the first Nissan Z to exceed the 250 km/h mark. Until 1989, this model was sold 3700 times in Germany. In the USA however, in 1984 alone, precisely 73,101 units went to the dealers. However, this car model had in the meantime received competition from the sports cars by Toyota, Mazda and Mitsubishi.

Model	Nissan 300 ZX T
Cubic capacity/cylinder	2960 cc/6 cylinder
hp/kW	240/176
Construction period	1985 – 1989
Number of produced cars	–

Nissan 300 ZX TT

In 1989, a Z model was introduced the likes of which had not been seen in the automotive world before. With twin turbo power and four-wheel steering, the 300 ZX entered the stage, a high technology vehicle that ventured into new areas. A V6 engine served by two turbochargers provided the race car, whose top speed was fixed at 250 km/h, with great power.

Model	Nissan 300 ZX TT
Cubic capacity/cylinder	2960 cc/6 cylinder
hp/kW	283/208
Construction period	1989 – 1996
Number of produced cars	–

Nissan 300 ZX TT

The chassis of 300 ZX TT with single-wheel suspension all around as well as 16-inch wheel rims corresponded absolutely to the state of the technology of its time. However, it was improved even further by the novelty of the four wheel steering system (!) developed by Nissan. Unfortunately, 1996 was the last year the 300 ZX TT was sold. Since that time, the fan community has been waiting for the revival of the Z legend.

Model	Nissan 300 ZX TT
Cubic capacity/cylinder	2960 cc/6 cylinder
hp/kW	283/208
Construction period	1989 – 1996
Number of produced cars	–

Nissan Silvia

The design of the first Silvia coupé goes back to the year 1964 and the drawing board of designer Albrecht Graf Goertz. The model became really famous in the 1980s when Nissan launched the fourth generation of this series. The modern stylish sports coupé with pop-up headlights was offered with several output levels and sold particularly well in the European market.

Model	Nissan Silvia
Cubic capacity/cylinder	1977 cc/4 cylinder
hp/kW	145/107
Construction period	1984 – 1989
Number of produced cars	–

Nissan 200 SX

As a successor of the Silvia coupé, Nissan presented in 1989 an improved version, the 200 SX. Under its bonnet, a turbo engine featured intercooling in addition to an electronic fuel injection system. A knock sensor controlled the combustion process in the engine. Unfortunately, the car was relatively expensive at a price of EUR 20,000.

Model	Nissan 200 SX
Cubic capacity/cylinder	1809 cc/4 cylinder
hp/kW	169/124
Construction period	1989 – 1994
Number of produced cars	–

Nissan 200 SX

When the second production series of the 200 SX was introduced in 1994, it was equipped as a novelty with a so-called multi-link suspension front and rear axle. In addition, the car was serially equipped with an ABS system and front air bags. Although the road performance of the 200 SX was considerable (maximum speed 235 km/h), this car led only a marginal existence in Europe.

Model	Nissan 200 SX
Cubic capacity/cylinder	1998 cc/4 cylinder
hp/kW	200/147
Construction period	1994 – 1999
Number of produced cars	–

Nissan 200 SX Racing

While the 200 SX developed into a best-seller in the Japanese and the US American markets, in Europe, and especially in Germany, it belonged to the difficult to market cars. Even a special model offered as of 1999 with sporty wide tyres and high-quality leather interior equipment did not improve the sales result.

Model	Nissan 200 SX Racing
Cubic capacity/cylinder	1998 cc/4 cylinder
hp/kW	200/147
Construction period	1999
Number of produced cars	800

Nissan Skyline GT R 32

The versions built from 1989 did not have anything in common anymore with the early Skyline of the 1960s. The once compact saloon car with the small 1.5-litre engine had developed into an aggressive sports car. Thanks to the strong turbo engine, the R 32 coupé was also used in its homeland for competitive racing where it scored one victory after another.

Model	Nissan Skyline GT R 32
Cubic capacity/cylinder	2595 cc/6 cylinder
hp/kW	280/206
Construction period	1989–1993
Number of produced cars	–

Nissan Skyline GT R 34

The skyline GT R 34 belonged, like its predecessor the R 32, to the cars which were aimed exclusively at the Japanese market. The just 460 centimetre long car with four-wheel drive was equipped with a six-cylinder engine. The car's output of 280 hp corresponded precisely to the upper limit stipulated by the Japanese authorities for a street sports car.

Model	Nissan Skyline GT R 34
Cubic capacity/cylinder	2595 cc/6 cylinder
hp/kW	280/206
Construction period	1993–1998
Number of produced cars	–

Nissan 100 NX GTI

In 1991, Nissan introduced the 100 NX, the coupé version of the Sunny model. Two years later the inconspicuous car also became available as a GTI version. With a maximum speed of 210 km/h, the GTI was based on a sports chassis and was serially equipped with rather generous features – sports seats, halogen headlights, painted fenders and much more were already included in the standard design version.

Model	Nissan 100 NX GTI
Cubic capacity/cylinder	1597 cc/4 cylinder
hp/kW	143/105
Construction period	1993 – 1995
Number of produced cars	–

Toyota Celica

When Toyota introduced the first Celica model in 1970, nobody could have foreseen that this car would constitute the foundation of a best selling series. This construction model persisted for more than three decades on the market. As part of ongoing model development, the Celica soon developed into an affordable middle class sports car.

Model	Toyota Celica
Cubic capacity/cylinder	1968 cc/4 cylinder
hp/kW	120/88
Construction period	1970 – 1978
Number of produced cars	–

Toyota Celica

When in 1977 the second production series of the Celica came off the conveyor belt, the shape of the coupé had already become a little closer to the prevailing European taste. The "baroque details" of its predecessor had to make room for modern styling elements like angular twin headlights and simple fenders. However, in spite of the bigger interior space, the car continued to be a 2 + 2-seater.

Model	Toyota Celica
Cubic capacity/cylinder	1588 cc/4 cylinder
hp/kW	75/55
Construction period	1977 – 1982
Number of produced cars	–

Toyota Celica

To be able to produce the Celica coupé as cost effectively as possible, Toyota used the underbody of the Carina saloon for this car. In terms of engine technology, both model series were identical – however, Celica customers understandably preferred the stronger aggregate. It provided a maximum speed of up to 195 km/h, in addition it could be easily tuned.

Model	Toyota Celica
Cubic capacity/cylinder	1968 cc/4 cylinder
hp/kW	123/90
Construction period	1977 – 1982
Number of produced cars	–

Toyota Celica

Under the motto "The shape of the future" the third Celica generation appeared in a rather wedge-shaped outfit that until that point was only associated with American cars. But under the metal exterior there was also a great amount of innovation – engines with output levels ranging from 86 hp to 124 hp, single-wheel suspension all around, as well as disc brakes for all wheels .

Model	Toyota Celica
Cubic capacity/cylinder	1587 cc/4 cylinder
hp/kW	124/91
Construction period	1982 – 1985
Number of produced cars	–

Toyota Celica Supra

The development engineers provided the top model of the third Celica generation with a six-cylinder-engine with twin camshafts propulsion. With this engine, the rear-wheel drive car theoretically advanced to the status of a competitor of the Porsche 924. However, due to its rather outdated chassis and the slightly longer wheel base, the Supra had to admit defeat.

Model	Toyota Celica Supra
Cubic capacity/cylinder	2759 cc/6 cylinder
hp/kW	170/125
Construction period	1982 – 1986
Number of produced cars	–

Toyota Celica

In 1985, with the launch of the fourth Celica generation, Toyota switched the propulsion of the successful coupé from rear to front-wheel drive. The four-cylinder engine (output between 116 hp and 185 hp) remained in use. However, it was now transversely installed. Depending on the provided features, the car with a top speed of 211 km/h cost between EUR 14,000 and EUR 17,500.

Model	Toyota Celica
Cubic capacity/cylinder	1587 cc/4 cylinder
hp/kW	150/110
Construction period	1985 – 1988
Number of produced cars	–

Toyota Celica

While the fifth Celica generation (1989 to 1993) was restyled externally with a primary focus on aerodynamics, Toyota decided as of 1994 to further update the image of the car. Toyota's American styling centre provided the sixth generation with another face-lift and equipped the sports coupé with wide tyres.

Model	Toyota Celica
Cubic capacity/cylinder	1998 cc/4 cylinder
hp/kW	175/129
Construction period	1994 – 1999
Number of produced cars	–

Toyota Celica Cabriolet

Unfortunately, not all Celica buyers could drive the car in the probably most enjoyable way – open. This was because the convertible version introduced at the Tokyo Motor Show in 1994 was intended only for the North American, Japanese, and Swiss markets. Due to the small number of produced cars, Toyota had the convertible once again built by ASC in the United States.

Model	Toyota Celica Cabriolet
Cubic capacity/cylinder	1998 cc/4 cylinder
hp/kW	175/129
Construction period	1994 – 1999
Number of produced cars	–

Toyota Celica

The specialist press was fairly astonished when the newest version of the Celica was presented to it in 1999 at the Frankfurt International Motor Show. According to the Toyota press release, the coupé with a top speed of 205 km/h consisted now of "… a model generation which was rolling directly into the future. (…) It is an all-round pleasure to look at, to drive and, of course, to own."

Model	Toyota Celica
Cubic capacity/cylinder	1794 cc/4 cylinder
hp/kW	143/105
Construction period	1999 – 2006
Number of produced cars	–

Toyota MR 2

In 1983, on the occasion of the 25th anniversary of the Tokio Motor Show, Toyota showed for the first time the prototype of a little two-seater sports car with removable roof halves. Within an astonishingly short time the concept car, which was still called SV-3 at that time, was ready to go into mass production. In 1985 already, the only four-meter long two-seater entered the market under the name MR 2.

Model	Toyota MR 2
Cubic capacity/cylinder	1587 cc/4 cylinder
hp/kW	124/91
Construction period	1985 – 1990
Number of produced cars	–

Toyota MR 2

The MR 2 was anything but a boring touring car. Those individuals, who put this model in their garages, were owners of a nice fun car with pop-up headlights which delivered one thing above all else – a large amount of driving pleasure. Incidentally, the Fiat group also featured in its program a similar vehicle concept with extremely wide C column and mid-engine – the model X1-9.

Model	Toyota MR 2
Cubic capacity/cylinder	1587 cc/4 cylinder
hp/kW	124/91
Construction period	1985 – 1990
Number of produced cars	–

Toyota MR 2

As opposed to the first production series, the second MR-2 generation manufactured from 1990 was substantially more rounded – the rough edges of its predecessor were a thing of the past, in this car the wind tunnel determined the external shape. The position of the four-cylinder engine was not changed, but the 1.6-litre engine was replaced by a two-litre version.

Model	Toyota MR 2
Cubic capacity/cylinder	1998 cc/4 cylinder
hp/kW	156/115
Construction period	1990 – 1994
Number of produced cars	–

Toyota MR 2

When Toyota equipped the MR 2 starting in 1994 with a 175-hp engine and allowed it to achieve a speed of 225 km/h, this was still far removed from the end of the development possibilities. There was an even stronger version in which a 225-hp turbo engine rumbled behind the seats; however, in the European market this version with the excessive engine, according to the opinion of experts, was not available.

Model	Toyota MR 2
Cubic capacity/cylinder	1998 cc/4 cylinder
hp/kW	175/129
Construction period	1994 – 1996
Number of produced cars	–

Toyota MR 2

The second generation MR 2 was able to assert its position in the market for almost one full decade. Although the car was far from being a best-seller, it underwent now and then some model development. Thus the cars destined for the USA received a solid coupé roof instead of the targa roof. The MR 2 models manufactured from 1997 had to make do with an especially large rear spoiler.

Model	Toyota MR 2
Cubic capacity/cylinder	1998 cc/4 cylinder
hp/kW	170/125
Construction period	1996 – 1999
Number of produced cars	–

Toyota Supra 3.0i

With the Supra, Toyota introduced a sports coupé under whose bonnet a strong six-cylinder-engine was at work. Upon its debut in 1986, the Supra was the most expensive Toyota model until that date – its starting price was EUR 25,000. In return for this amount, buyers received a car with a top speed of 220 km/h whose basic equipment included an electrically adjustable driver seat, a cruise control and much more.

Model	Toyota Supra 3.0i
Cubic capacity/cylinder	2954 cc/6 cylinder
hp/kW	204/150
Construction period	1986 – 1988
Number of produced cars	–

Toyota Supra 3.0i

In the second production series manufactured from 1993, the Supra became even more sophisticated, because it had matured into a full-fledged sports car, which moved on 17 inch wheels and featured, of course, a gigantic back spoiler. With this outfit it corresponded fully to the preferences of its time. The model with a top speed of 250 km/h, required a mere 5.1 seconds to accelerate to the 100 km/h mark.

Model	Toyota Supra 3.0i
Cubic capacity/cylinder	2997 cc/6 cylinder
hp/kW	330/243
Construction period	1993 – 1997
Number of produced cars	–

Toyota Supra Turbo

After Toyota stopped exporting the Supra in 1997, it continued to produce this model for the Japanese market for two more years. In Japan, the top version with a turbo charger cost about EUR 52,000. To comply with the wishes of the Japanese customers, Toyota offered the Supra alternatively also without the back spoiler. However, a retroactive refitting was not possible.

Model	Toyota Supra Turbo
Cubic capacity/cylinder	2997 cc/6 cylinder
hp/kW	330/243
Construction period	1997 – 1999
Number of produced cars	–

Toyota Paseo

The Paseo, a so-called 2+2-seater, was not a Japanese, but an American design. It originated on the drawing board of the California Toyota design centre. In the United States the Paseo was offered already as of 1991. After an optical update the model also appeared on the European market five years later.

Model	Toyota Paseo
Cubic capacity/cylinder	1497 cc/4 cylinder
hp/kW	90/66
Construction period	1996 – 1999
Number of produced cars	–

Toyota Paseo Cabriolet

At the Geneva Motor Show in 1997, Toyota presented the Paseo – which was marketed in Japan under the model name Cynos – for the first time as a convertible version. Because the convertible was manufactured only in small production numbers, Toyota commissioned the American company ASC (American Sunroof Cabriolet) to carry out the conversion.

Model	Toyota Paseo Cabriolet
Cubic capacity/cylinder	1497 cc/4 cylinder
hp/kW	90/66
Construction period	1997 – 1999
Number of produced cars	–

2000
today

The fastest, most expensive and wildest models

Car brands and their image

Since the turn of the millennium, it seems that the times when the production lines of the automobile manufacturers produced unemotional standard sports cars are finally a thing of the past. More and more elegant roadsters, convertibles, coupés and sports saloons compete for the favour of affluent buyers. In addition, one thing seems for sure: there will be an unprecedented large number of extremely fast sports cars introduced in the next few years. Following decades in which sceptics and appointed experts already prophesied the definitive end of expensive sports cars, we are facing the most extensive flood of high horse power and luxurious cars than ever before.

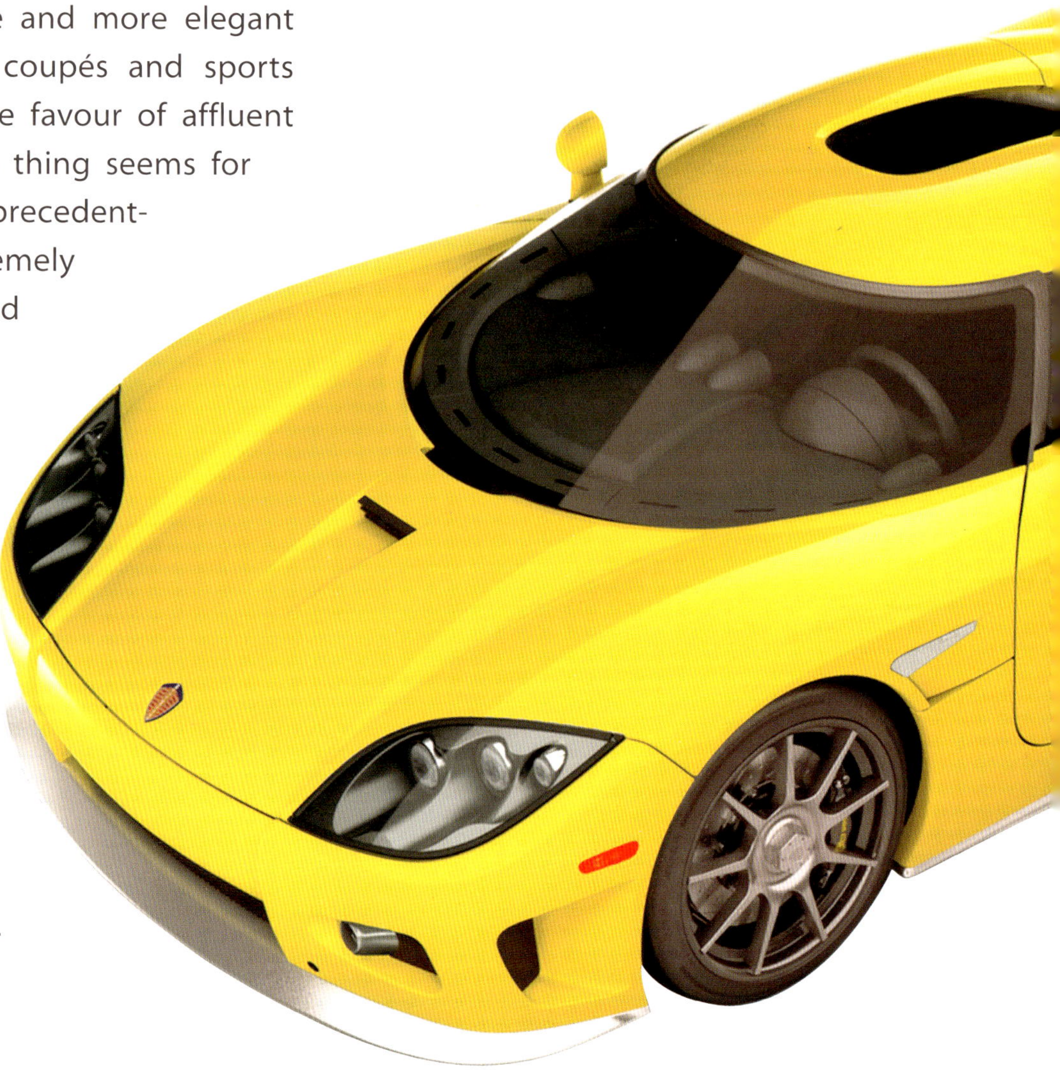

Legendary models of the past got very modern successors (for example, Alfa Romeo 8C) or at least preliminarily followed the footprints of their predecessors – a famous name is an obligation to deliver. The manufacturers are very busy developing as much as they can to present quite early the first studies that allow us to foresee the design of the next vehicle generation. However, the serial models usually do not turn out to be quite as futuristic as proclaimed – even if the manufacturers try to sell some technical concept or another as an absolute novelty.

For example, similar to many other features, the so-called Vario collapsible hood – a construction between the classical soft top and a hard top – already existed in the 1930s.

However, what is new is the fact that since the year 2000 high performance sports cars increasingly benefitted from the most advanced material technology. Examples of this new trend include car bodies that are made increasingly out of carbon fibre materials and brake systems that are

equipped with ceramic brake discs. In addition, more and more often manufacturers consider dynamic chassis settings to be standard. In this aspect, no doubt the Formula One is the big role model – as not only in terms of technology, but also in terms of appearance there are a number of racers that are modelled on racing sports cars. It seems that the long since outdated advertising slogan of the 1960s "Only flying is better", will probably soon become a reality.

Anyone, who carefully examined the exhibits at the big international motor shows of the past few years, may have noted one or the other newcomer among the automobile manufacturers. There is no specialised periodical that has not reported in the past about noble brands like Fisker or Pagani. Even less known names can be found at the Millionaire Fair taking place in Moscow or Monte Carlo's Top Marques Luxury Fair. The advantage for these exhibitors who annually only produce (mostly manually) a few dozen noble vehicles in exhibiting their products at such fairs is obvious. It is a venue where they can meet face to face with the affluent customers who are actually waiting to purchase a new and exotic vehicle (maybe a Panoz or Tramontana?) – provided that the car has plenty of cylinders and offers an abundance of performance.

The sports car manufacturers have no problems nowadays in fulfilling the supposed demand of customers for more and more engine power. In almost quarterly intervals they add a few hp here and there. This act seems to be very important for image reasons mainly – such tricks do not influence the maximum velocity of a car in any significant way.

When the first lines of this book were written, the ultimately achievable power was 1001 hp (Bugatti Veyron). Shortly before going to press, there were first rumours in the media that a car manufacturer from the USA was planning to launch in 2008 the world's most exclusive car. The car, with a selling price of about two million dollars, will allegedly dwarf all vehicles ever built before, because it will be powered by a 16-cylinder engine with 14 litres of cubic capacity – while the power output shall amount to 1200 hp!

Audi Study Rosemeyer

The Audi "Rosemeyer" design study is not a ready to drive prototype. However, it could be powered – like the legendary Silver Arrow racers – by a 16 cylinder mid-engine. The appearance of the study with the monumental radiator grille deliberately resembles the historical Auto Union Silver Arrows, and similar to them, the aluminium body reflects the light in a way that creates a magically shining aura.

Model	Audi Study Rosemeyer
Cubic capacity/cylinder	–
hp/kW	–
Construction period	2000
Number of produced cars	unique piece

Audi Nuvolari Quattro

Audi presented the draft study "Nuvolari Quattro" at the Geneva International Motor Show. The design of the Nuvolari Quattro consists of a 2+2-seater coupé that exhibits the proportions of a classical GT. It is adequately powered by the strongest engine ever developed by Audi for the application in a street vehicle – a 5.0-litre V10-biturbo charging engine with FSI direct fuel injection.

Model	Audi Nuvolari Quattro
Cubic capacity/cylinder	5000 cc/10 cylinder
hp/kW	600/441
Construction period	2003
Number of produced cars	study

Audi Le Mans Quattro

The Audi Le Mans Quattro is the study of a top performance sports car, which will most probably go into production as of 2007. It has a width of 190 centimetres, a length of only 437 centimetres and a height of 125 centimetres. Its wheel base of 265 centimetres allows the construction of a spacious cockpit, so that nothing stands in the way of sporty yet comfortable driving pleasure.

Model	Audi Le Mans Quattro
Cubic capacity/cylinder	5000 cc/10 cylinder
hp/kW	610/449
Construction period	2003
Number of produced cars	study

BMW Z3 3.0i

In addition to the four- and six-cylinder versions, the Z3 was also offered in two "M" versions: the M Roadster and M Coupé profited from the 3.2-litre high-capacity engine of the BMW M3 (325 hp). The result of this combination is the fact that to this day, drivers always prefer to close the door of this sports car from the inside, because no other car embodies the philosophy of BMW as fully as the Z3: "Sheer driving pleasure".

Model	BMW Z3 3.0i
Cubic capacity/cylinder	2979 cc/6 cylinder
hp/kW	231/170
Construction period	2000 – 2003
Number of produced cars	–

BMW Z4 2.5i

The BMW Z4, which was already available on the US market in late 2002, is exclusively produced at the American BMW Spartanburg factory (4000 employees) for global export. Because the Z4 is a very complicated vehicle that consists of substantially more parts than its predecessor, the construction of the bodywork alone requires a production area of 19,000 square metres.

Model	BMW Z4 2.5i
Cubic capacity/cylinder	2494 cc/6 cylinder
hp/kW	192/141
Construction period	from 2002
Number of produced cars	

BMW Z4 3.0i

Upon request, the BMW Z4 is available with a fully automated convertible roof – the electromechanical process requires ten seconds at the touch of a button. Starting in autumn 2003 a hard top also became available. Its distinguishing feature is the cross flow spoiler in the rear roof area, which strengthens the drift and thereby improves the handling characteristics.

Model	BMW Z4 3.0i
Cubic capacity/cylinder	2979 cc/6 cylinder
hp/kW	231/170
Construction period	from 2002
Number of produced cars	–

BMW Z4 M Roadster

The Z4 M Roadster is equipped with a speed-sensitive variable differential lock – which helps experienced drivers to strengthen the positive qualities of the rear-wheel drive in sporty driving. In difficult weather conditions, which occur primarily in winter, this intelligent system gives the Z4 M Roadster especially strong ground traction.

Model	BMW Z4 M Roadster
Cubic capacity/cylinder	3246 cc/6 cylinder
hp/kW	343/252
Construction period	from 2006
Number of produced cars	–

BMW Z4 Coupé Study

In 2005, BMW introduced the sensational Z4 coupé concept car. The matte deluxe painted body, designed in the style of a classical sports coupé, thrilled the visitors of the Frankfurt International Motor Show as well as the visitors of the Tokyo Fair shortly thereafter. Due to the overwhelmingly positive feedback, it was decided to force the serial development of the coupé – in June 2006, the first cars were delivered to the customers!

Model	BMW Z4 Coupé Study
Cubic capacity/cylinder	2996 cc/6 cylinder
hp/kW	265/195
Construction period	2005
Number of produced cars	unique piece

BMW Z4 Coupé

A strong straight six-cylinder engine (265 hp/195 kW) provides the Z4 Coupé not only with agility, but by the use of magnesium also with a lower weight. The high-end version of this sports car is called, incidentally, Z4 M Coupé. It is powered by a 343 hp (252 kW) high revolution engine – thus equipped, the car reaches a maximum speed of 250 km/h.

Model	BMW Z4 Coupé
Cubic capacity/cylinder	2996 cc/6 cylinder
hp/kW	265/195
Construction period	from 2006
Number of produced cars	–

BMW Z8

With the two-seater sports car Z8, BMW has certainly succeeded in creating a modern interpretation of the classic 507 model. Just like the 1950s original, the Z8 also thrills with a harmoniously proportioned perfect figure: 440 centimetres long, 183 centimetres wide and 131 centimetres high. The precursor of the Z8 was a study, which was already shown in 1997 under the name Z07.

Model	BMW Z8
Cubic capacity/cylinder	4941 cc/8 cylinder
hp/kW	400/294
Construction period	2000 – 2003
Number of produced cars	5700

BMW Z8

The "face" of the Z8 is unmistakably that of a BMW. Its wide kidney grille as well as small Xenon round headlights underline the traditional brand characteristics. Modern aluminium technology provides Z8 owners with a car body whose value remains constant for a long time because this type of construction is absolutely rust-proof. The Z8 reaches the 100 km/h mark after 4.7 seconds, its maximum speed is 250 km/h.

Model	BMW Z8
Cubic capacity/cylinder	4941 cc/8 cylinder
hp/kW	400/294
Construction period	2000 – 2003
Number of produced cars	5700

BMW M3

An M3 cost DM 58.000 in the year of its market launch in 1986. Yet it was no problem to find customers for the car even in this price range. To this day, this fact has not changed and the fascination emitted by the M3 remains unbroken. For comparison, the basic version of the M3 coupé built in 2000 cost about DM 100.000.

Model	BMW M3
Cubic capacity/cylinder	3246 cc/6 cylinder
hp/kW	343/252
Construction period	2000 – 2006
Number of produced cars	

BMW M3 Cabrio

While the M3 Cabrio had to undergo a face-lift for the year 2001, it was not negatively affected by the change – the tightened waistline made the car with a top speed of 250 km/h seem even stronger and more athletic. The torque of the six-cylinder engine was transmitted, incidentally, by a six-step gear to the wide wheels (front 225/45 ZR 18 and rear 255/40 ZR 18).

Model	BMW M3 Cabrio
Cubic capacity/cylinder	3246 cc/6 cylinder
hp/kW	343/252
Construction period	from 2001
Number of produced cars	–

BMW M3 CSL

With the M3 CSL presented at the Geneva International Motor Show in 2003, the BMW M GmbH has introduced a powerful car that revives the era of lightweight construction coupés. The concept of the CSL is reminiscent in all aspects of a classical competition touring car, which is why the M3 produced in a small volume is almost predestined for competitive racing.

Model	BMW M3 CSL
Cubic capacity/cylinder	3246 cc/6 cylinder
hp/kW	360/265
Construction period	from 2003
Number of produced cars	–

BMW M5

For a long time, the M5 has been exploring levels of dynamic driving qualities that go far beyond the scope of the conventional segment of sports saloons. For example, the current model presented in March 2004 reaches the 0 to 100 km/h mark after 4.7 seconds and achieves a regulated maximum speed of 250 km/h in the seventh gear – while today there is a V10 cylinder under its bonnet!

Model	BMW M5
Cubic capacity/cylinder	4999 cc/10 cylinder
hp/kW	507/373
Construction period	from 2004
Number of produced cars	–

BMW M6

With the new M6, the BMW M GmbH presents the 6-series deluxe coupé in its sportiest form. This M6 is the noblest and strongest car ever developed for the 6-series – with five litres of cubic capacity and ten cylinders! As opposed to its competitors that generally consist of two-seaters, the M6 offers the comfort of a 2+2-seater coupled with the luxurious equipment of a typical BMW of the luxury class.

Model	BMW M6
Cubic capacity/cylinder	4999 cc/10 cylinder
hp/kW	507/373
Construction period	from 2004
Number of produced cars	–

BMW M6

BMW M has always been a synonym for sporty cars with the highest output. The new BMW M6 continues this tradition and sets a new yardstick in this segment with its combination of sportiness and elegance. Actually, there is no direct predecessor of the M6 – many of its details (ten-cylinder engine, seven-step gears etc.) were never introduced in earlier models.

Model	BMW M6
Cubic capacity/cylinder	4999 cc/10 cylinder
hp/kW	507/373
Construction period	from 2004
Number of produced cars	–

BMW 645 Ci Coupé

With its elongated proportions, the new car of the 6-series belongs to the category of the sportily elegant coupés. Its dynamic appearance already indicates its special driving agility – In conjunction with modern driving assistance systems like the dynamic drive or the driving stability controller DSC, it provides an impressive driving experience.

Model	BMW 645 Ci Coupé
Cubic capacity/cylinder	4398 cc/8 cylinder
hp/kW	333/245
Construction period	from 2003
Number of produced cars	–

BMW 645 Ci Cabriolet

Shortly after the 645 Ci Coupé celebrated its debut at the Frankfurt International Motor Show in 2003, BMW presented in the USA at the Detroit Motor Show another highlight – the open variation of this dream car! The 2 + 2-seater 645 Ci Cabriolet, which closely resembles the 645 Ci Coupé, was put on the market for the first time in Spring 2004 for a starting price of around EUR 80,000.

Model	BMW 645 Ci Cabriolet
Cubic capacity/cylinder	4398 cc/8 cylinder
hp/kW	333/245
Construction period	from 2004
Number of produced cars	–

Gumpert Apollo

Roland Gumpert, engineer by profession and former head of the Audi rally department, is among the few individuals who have had the chance to build their own car. His Apollo presented in 2005 is a purist vehicle that offers an incomparable driving experience on the street coupled with the potential of achieving victories on the racetrack as well.

Model	Gumpert Apollo
Cubic capacity/cylinder	4163 cc/8 cylinder
hp/kW	650/478
Construction period	from 2005
Number of produced cars	–

Gumpert Apollo

Dynamism and sportiness distinguish the design and technology of the only 110 centimetre-high Apollo. The turbo-loaded V8 engine provides abundant strength combined with breathtaking acceleration values (from 0 to 100 km/h in 3 seconds). A taut chassis and the ingenious aerodynamics of the body guarantee controlled vehicle handling even when going at the maximum speed (360 km/h).

Model	Gumpert Apollo
Cubic capacity/cylinder	4163 cc/8 cylinder
hp/kW	650/478
Construction period	from 2005
Number of produced cars	–

Maybach Exelero

It sounds incredible but is actually true: to be able to demonstrate the performance potential of its tyres, the tyre manufacturer Fulda commissioned Maybach to develop this exclusively unique car. It is obvious that the design of the Maybach Exelero is strongly reminiscent of the late 1930s, when Maybach cars (just like today) belonged to the most luxurious vehicles available on the market.

Model	Maybach Exelero
Cubic capacity/cylinder	5908 cc/12 cylinder
hp/kW	700/515
Construction period	2005
Number of produced cars	unique piece

Maybach Exelero

The Exelero, a fully roadworthy unique car with superlative qualities, posed especially high challenges for its design engineers. The maximum speed of the nearly six metre long deluxe coupé was required to amount to 350 km/h. Test measurements have confirmed this value – the Exelero – which moves on Fulda Carat Exelero tyres measuring 315/25 ZR 23 – can reach precisely 351.45 km/h!

Model	Maybach Exelero
Cubic capacity/cylinder	5908 cc/12 cylinder
hp/kW	700/515
Construction period	2005
Number of produced cars	unique piece

Mercedes-Benz SLA Study

The SLA study is a small roadster on the basis of the A class with accordingly compact dimensions – its length of 377 centimetres promises an agile driving experience. Powerfully arched fenders, a level windshield, big doors and a softly tapered rear in the style of the legendary Mercedes Silver Arrows could be the distinguishing features of the roadster design of the future.

Model	Mercedes-Benz SLA Study
Cubic capacity/cylinder	1900 cc/4 cylinder
hp/kW	125/92
Construction period	2000
Number of produced cars	unique piece

Mercedes-Benz SLK 320

Nowadays the SLK has attained the image of a modern cult car, which is impressively confirmed by the great demand for this model in the used car market. More than 308,000 owners in all corners of the globe, more than 40 international awards and the global top rank of its market segment – this is the considerable sum of achievemenst of this first generation of the SLK class.

Model	Mercedes-Benz SLK 320
Cubic capacity/cylinder	3199 cc/6 cylinder
hp/kW	218/160
Construction period	2000 – 2003
Number of produced cars	–

Mercedes-Benz SLK 32 AMG

Mercedes AMG introduced the Roadster SLK 32 AMG in 2001. The road performance and the sporty chassis are great fun, because its V6-compressor engine is a technical masterpiece. The acoustic of the engine is also distinguished by careful attention to technological details – the sound of the machine offers quite a sound experience without any annoying side effects.

Model	Mercedes-Benz SLK 32 AMG
Cubic capacity/cylinder	3199 cc/6 cylinder
hp/kW	354/260
Construction period	2001 – 2003
Number of produced cars	–

Mercedes-Benz SLK 200 K

Eight years after the première of the first SLK, the second SLK generation was introduced in the spring of 2004. The proven strengths of the predecessor had been further developed – above all with a new, space-saving retractable hardtop, a more innovative security system and other new developments like the bi-xenon headlights with cornering assist.

Model	Mercedes-Benz SLK 200 K
Cubic capacity/cylinder	1796 cc/4 cylinder
hp/kW	163/120
Construction period	from 2004
Number of produced cars	–

Mercedes-Benz SLK 350

The SLK class is available in three model variations: the four-cylinder SLK 200 compressor, the SLK 350 with the newly developed V6 engine and – for the first time in this vehicle class – an eight-cylinder engine by Mercedes-AMG. All SLK models feature dynamical shapes – with beautifully designed details they emphasise their origin from a manufacturer with an extensive roadster tradition.

Model	Mercedes-Benz SLK 350
Cubic capacity/cylinder	3498 cc/6 cylinder
hp/kW	272/200
Construction period	from 2004
Number of produced cars	–

Mercedes-Benz SLK 55 AMG

Those who consider the output and prestige of the eight- and twelve-cylinder models not sufficient, can be served by "AMG" – the sports department of the brand. Since 1999, AMG has been called "Mercedes-AMG GmbH" and is a subsidiary of the Daimler AG. It creates especially powerful and agile car variations which allow the acceleration from 0 to 100 km/h in an extremely short time.

Model	Mercedes-Benz SLK 55 AMG
Cubic capacity/cylinder	5439 cc/8 cylinder
hp/kW	360/265
Construction period	from 2004
Number of produced cars	

Mercedes-Benz SL 500

The fifth SL generation (R 230) introduced in summer 2001, is once more a car whose technical innovations are as futuristic as those of the 300 SL gull-wing door model of 1954. With its electronic chassis system, this SL offers a maximum of driving security.

Model	Mercedes-Benz SL 500
Cubic capacity/cylinder	5461 cc/8 cylinder
hp/kW	388/285
Construction period	from 2001
Number of produced cars	–

Mercedes-Benz SL 350

As opposed to its predecessors, since the year 2001 the SL always carries its hard top along with it. At the touch of a button the top can be opened or closed within 16 seconds. In order to save space, the ingenious swivel mechanism allows the three roof sections to completely disappear into the upper area of the boot – while keeping a boot space of 235 litres available.

Model	Mercedes-Benz SL 350
Cubic capacity/cylinder	3498 cc/6 cylinder
hp/kW	272/200
Construction period	from 2002
Number of produced cars	

Mercedes-Benz SL 600

The powerful engines, which are among the most modern of their class, guarantee an unusual driving experience. In addition, they considerably underbid the strict exhaust limits of the EU-4 norm. The V12 engine of the SL 600 accelerates the roadster, which weighs almost two metric tons, within 4.5 seconds from 0 to 100 km/h – while its maximum speed is 250 km/h.

Model	Mercedes-Benz SL 600
Cubic capacity/cylinder	5514 cc/12 cylinder
hp/kW	517/380
Construction period	from 2003
Number of produced cars	–

Mercedes-Benz SL 55 AMG

In terms of appearance, the AMG Roadster introduced at the Frankfurt International Motor Show in 2001 differs from the SL 500 by more pronounced bumpers and side skirts. Under its bonnet works a 5.5-litre engine that accelerates the car from 0 to 100 km/h in 4.7 seconds – while the electronically restricted maximum speed is 250 km/h.

Model	Mercedes-Benz SL 55 AMG
Cubic capacity/cylinder	5439 cc/8 cylinder
hp/kW	500/368
Construction period	2001 – 2005
Number of produced cars	–

Mercedes SL 55 AMG

From 2006, the revised engine of the SL 55 AMG, featuring a new compressor with higher revolution, provided the car with even more power. In conjunction with the application of a more powerful engine control device, the output of the AMG-V8 engine could thus be increased from 500 hp to 517 hp. Now the acceleration from 0 to 100 km/h takes 4.5 seconds – previously it took 4.7 seconds.

Model	Mercedes-Benz SL 55 AMG
Cubic capacity/cylinder	5439 cc/8 cylinder
hp/kW	517/380
Construction period	from 2006
Number of produced cars	–

Mercedes-Benz SL 65 AMG

The SL 65 AMG which is available starting at about EUR 206,100, celebrated its world première at the Geneva International Motor Show in 2006. It is powered by an AMG six-litre V12 biturbo engine and its 612 hp allows it to accelerate in 4.2 seconds from a standing position to 100 km/h. The maximum speed is electronically limited to 250 km/h.

Model	Mercedes-Benz SL 65 AMG
Cubic capacity/cylinder	5980 cc/12 cylinder
hp/kW	612/450
Construction period	from 2006
Number of produced cars	–

Mercedes-Benz SLR McLaren

With the McLaren SLR the Stuttgart automobile manufacturers and their Formula One partner McLaren documented their long-standing experience in the development and production of top output sports car. The two-seater with the striking gull wing doors carries on the myth of the legendary SLR racing sports cars from the 1950s.

Model	Mercedes-Benz SLR McLaren
Cubic capacity/cylinder	5439 cc/8 cylinder
hp/kW	626/460
Construction period	from 2003
Number of produced cars	–

Mercedes-Benz SLR McLaren

The newly developed V8-compressor engine of the SLR accelerates the sports car from 0 to 100 km/h in 3.8 seconds. The maximum speed amounts to 334 km/h. Similar to the Formula One racers of McLaren Mercedes, the bodywork of the SLR also consists of the car bon fibre materials which offer an exemplary energy absorption coupled with low weight and thus the highest passenger's security.

Model	Mercedes-Benz SLR McLaren
Cubic capacity/cylinder	5439 cc/8 cylinder
hp/kW	626/460
Construction period	from 2003
Number of produced cars	–

Mercedes-Benz SLR 722 Edition

With the model variation SLR "722 Edition" Mercedes-Benz commemorates the unforgotten victory which Stirling Moss and his Mercedes-Benz 300 SLR gained with the start number 722 in 1955. The special edition model is manually assembled by AMG. It accelerates from 0 to 100 km/h in 3.6 seconds and has a top speed of 337 km/h.

Model	Mercedes-Benz SLR 722 Edition
Cubic capacity/cylinder	5500 cc/8 cylinder
hp/kW	650/478
Construction period	from 2006
Number of produced cars	–

Opel Speedster

When the light Opel Speedster weighing only 850 kilograms was launched in the beginning of 2001 on the European market, it was initially powered by a 2.2-litre four-cylinder engine – enough power for a maximum speed of 220 km/h. The two-seater with a mid-engine and aluminium chassis went from 0 to 100 km/h in less than six seconds.

Model	Opel Speedster
Cubic capacity/cylinder	2200 cc/4 cylinder
hp/kW	147/108
Construction period	2001 – 2005
Number of produced cars	7000

Opel Speedster Turbo

As of 2003, Opel offered Speedster fans who wanted an especially sportive drive the "Speedster Turbo" version. Despite its low noise level, its road performance was nevertheless absolutely at sports car level. Its turbo power catapulted the compact and extremely low riding two-seater in 4.9 seconds to a speed of 100 km/h.

Model	Opel Speedster Turbo
Cubic capacity/cylinder	2200 cc/4 cylinder
hp/kW	200/147
Construction period	2003 – 2005
Number of produced cars	–

Opel Speedster

After a construction period of almost five years the last version of the Speedster came off the conveyor belt in July 2005. The purist roadster made Opel history. In co-operation with Lotus, this was the first time that a sporty Opel with a mid-engine had been developed. Aluminium and plastic were used as materials for the chassis and bodywork.

Model	Opel Speedster
Cubic capacity/cylinder	2200 cc/4 cylinder
hp/kW	147/108
Construction period	2001 – 2005
Number of produced cars	7000

Opel Astra Coupé OPC X-Treme Concept Car

An Opel super sports car study called "Astra Coupé OPC X-Treme" celebrated its world première at the Geneva International Motor Show in 2001. The design of the roadworthy two-seater sports car was based on the Astra-Coupé V8 racing car that was successfully entered into the Deutsche Tourenwagen Masters (DTM, German Touring Car Masters).

Model	Opel Astra Coupé OPC X-Treme Concept Car
Cubic capacity/cylinder	4000 cc/8 cylinder
hp/kW	444/326
Construction period	2001
Number of produced cars	unique piece

Opel Astra Coupé OPC X-Treme Concept Car

The visual eye-catcher of the red painted Astra Coupé OPC X-Treme (length/width/height: 4290/1850/1290 millimetres) were its two gull-wing doors which gently floated upwards, controlled by gas pressure shock absorbers. Special 20-inch tyres in the size 245/25 in front and 285/25 in the rear underlined the powerful appearance of this concept car.

Model	Opel Astra Coupé OPC X-Treme Concept Car
Cubic capacity/cylinder	4000 cc/8 cylinder
hp/kW	444/326
Construction period	2001
Number of produced cars	unique piece

Opel Vectra GTC OPC

In terms of appearance alone, the Vectra GTC OPC is an eye-catcher – large cooling air intakes and a honeycomb-structured front grille, 18-inch wheels as well as two trapezoid-shaped exhaust tailpipes underline the sporty character of this model. With the support of a turbo, its maximum speed is 260 km/h. The starting price of the strong car amounts to about EUR 37,000.

Model	Opel Vectra GTC OPC
Cubic capacity/cylinder	2792 cc/6 cylinder
hp/kW	255/188
Construction period	from 2005
Number of produced cars	–

Opel GT

With a sharp design, rear-wheel drive and a turbo engine, the new Opel GT continues as of 2007 the successful concept of the first GT, which Opel launched in 1968. However, the new GT is not a coupé but a roadster. With its soft top, it offers an open air experience that was denied the owners of the first GT.

Model	Opel GT
Cubic capacity/cylinder	1998 cc/4 cylinder
hp/kW	260/191
Construction period	from 2007
Number of produced cars	–

Porsche 911 Turbo Cabriolet

A new highlight in the world of convertibles was introduced in 2003 – ample air intakes in the front and sides indicate that after 14 years an open Porsche Turbo is available once again. Two superchargers serve the engine of the sports car – therefore the acceleration to the 100 km/h mark takes a mere 4.3 seconds. It reaches the 160 km/ h mark within 9.5 seconds, and its forward push only ends at 305 km/h.

Model	Porsche 911 Turbo Cabriolet
Cubic capacity/cylinder	3600 cc/6 cylinder
hp/kW	420/309
Construction period	from 2003
Number of produced cars	–

Opel GT

The development history of the new Opel GT is a classic example of the collaboration in the General Motors Group across continents. For example, its design was inspired by the concept car "Vauxhall VX Lightning" which originated in 2003 at the GM "Advanced Design Studio" in Birmingham/England. The new Opel GT, however, is produced in Wilmington in the US State of Delaware.

Model	Opel GT
Cubic capacity/cylinder	1998 cc/4 cylinder
hp/kW	260/191
Construction period	from 2007
Number of produced cars	–

Porsche 911 Turbo Cabriolet

Like the 911 coupé, the new 911 Turbo Cabriolet is serially equipped with four-wheel drive via a visco lamella clutch. On non-slip roads, five percent of the drive torque is assigned to the front. If the state of the road surface deteriorates, becoming wet or slick for instance, up to 40 percent of the torque can be applied to the front wheels.

Model	Porsche 911 Turbo Cabriolet
Cubic capacity/cylinder	3600 cc/6 cylinder
hp/kW	420/309
Construction period	from 2003
Number of produced cars	–

Porsche 911 "40 Jahre 911" Special Model

In 2003, on the occasion of the 40th birthday of the 911 model, Porsche introduced a jubilee edition limited to 1963 cars. The "40 Jahre 911" (40 years of 911) special model only available painted in "silver-metallic GT", was at the same time a climax of the current "996" series. It was tuned for an especially sporty performance and luxuriously equipped.

Model	Porsche 911 "40 Jahre 911"
Cubic capacity/cylinder	3600 cc/6 cylinder
hp/kW	345/254
Construction period	2003
Number of produced cars	1963

Porsche 911 "40 Jahre 911" Special Model

The Porsche 911 Carrera constituted the base of the "40 Jahre 911" special model that cost around EUR 96,000. Porsche offered the special model with a top speed of 290 km/h only in "silver metallic GT", a colour that is normally reserved for the Carrera GT. The various distinguishing features of the jubilee edition included polished 18-inch Carrera wheels.

Model	Porsche 911 "40 Jahre 911"
Cubic capacity/cylinder	3600 cc/6 cylinder
hp/kW	345/254
Construction period	2003
Number of produced cars	1963

Porsche 911 GT 2

The Porsche GT 2 (996 series) with a top speed of 315 km/h constitutes the upper end of the series in terms of output. The engine of this model built for the use on regular streets is based on the 911 GT 1 engine, which has been tested in racing. The first GT 2 (420 hp) of the year 2000 was followed by a revised version with even more bite three years later.

Model	Porsche 911 GT 2
Cubic capacity/cylinder	3600 cc/6 cylinder
hp/kW	462/340
Construction period	from 2003
Number of produced cars	–

Porsche 911 GT 3

More achievement with the same cubic capacity and the same fuel consumption – this was the formula for developing the 911 GT 3 of the second generation. The GT 3, which was added to the brand's model range in spring 2003, is the condensed outcome of half a century of Porsche car racing experience. A pure sports car – with all the classical virtues of this type of vehicle and without anything that could mar the pure driving pleasure.

Model	Porsche 911 GT 3
Cubic capacity/cylinder	3600 cc/6 cylinder
hp/kW	381/280
Construction period	from 2003
Number of produced cars	–

Porsche 911 GT 3

By reducing the mass of the moving engine parts, the 911 GT disposes of an output of 360 hp and accelerates in 4.5 seconds to a speed of 100 km/h. Upon request, the GT 3 can also be equipped with a Porsche ceramic composite brake. These ceramic composite brake discs with a diameter of 350 millimetres weigh 50 percent less than metal brake discs.

Model	Porsche 911 GT 3
Cubic capacity/cylinder	3600 cc/6 cylinder
hp/kW	381/280
Construction period	from 2003
Number of produced cars	–

Porsche 911 GT 3 RS

In 2003, Porsche continued is model launch with another 911 version called GT 3 RS – with the abbreviation "RS" being especially exciting for racing fans. At a cost of approximately EUR 120,800, the RS has a maximum speed of 306 km/h and requires 4.4 seconds for the sprint from 0 to 100 km/h. Porsche plans to build the car that is normally painted white (featuring the logotype in red or blue) at least 200 times.

Model	Porsche 911 GT 3 RS
Cubic capacity/cylinder	3596 cc/6 cylinder
hp/kW	381/280
Construction period	from 2003
Number of produced cars	–

Porsche Carrera GT

The appearance of the Carrera GT already reflects its performance capacity as an uncompromising super sports car. However, as opposed to racing concept cars, the design takes into consideration the stylistic relationship to serial vehicles and is reminiscent in certain details of the legendary Porsche racing cars. Thus the typical Porsche face is that of the 718 RS Spyder of the 1960s.

Model	Porsche Carrera GT
Cubic capacity/cylinder	5733 cc/V10 cylinder
hp/kW	612/450
Construction period	2003
Number of produced cars	–

Porsche Carrera GT

Seen laterally, the 461 centimetre long, 192 centimetre wide and 116 centimetre high Carrera GT looks very dynamic due to its compact shoulder shape. The whole design of the car with a maximum speed of 330 km/h (acceleration from 0 to 100 km/h in 3.9 seconds) is influenced by its aerodynamics. A spoiler automatically extends at a speed of 120 km/h and provides the necessary output for the rear axle.

Model	Porsche Carrera GT
Cubic capacity/cylinder	5733 cc/V10 cylinder
hp/kW	612/450
Construction period	2003
Number of produced cars	–

Porsche Boxster „50 Jahre 550 Spyder" Anniversary Model

In January 2004, Porsche presented an interesting Boxster special model in commemoration of the 50th anniversary of the legendary Type 550 Spyder of the year 1953. As appropriate for the event, the series was limited to 1953 units. In addition, it was equipped with many details that so far had not been available in the Boxster – including a significantly shortened shifting travel of the gearshift lever of the car whose top speed reached 266 km/h.

Model	Porsche Boxster "50 Jahre 550 Spyder" Anniversary Model
Cubic capacity/cylinder	3200 cc/6 cylinder
hp/kW	266/196
Construction period	2004
Number of produced cars	1953

Porsche Cayman S

The Cayman S mid-engine coupé introduced by Porsche in 2005 was chosen in the meantime in New York as the "World Performance Car" of 2006. The Cayman S beat exactly 29 competitors to this significant award! The model with a top speed of 275 km/h is technically located between the Porsche Boxster and the 911 Carrera – the driving pleasure is available starting at approximately EUR 58,600.

Model	Porsche Cayman S
Cubic capacity/cylinder	3386 cc/6 cylinder
hp/kW	295/217
Construction period	from 2005
Number of produced cars	–

Porsche Cayman

After the successful market launch of the Cayman S in autumn 2005, Porsche in 2006 introduced the Cayman, another variation of the mid-engine-sports coupé. While its maximum speed is 258 km/h, the sports car accelerates from 0 to 100 km/h in 6.1 seconds. The Cayman moves in thefront on tyres sized 205/55, and in the rear on tyres sized 235/50.

Model	Porsche Cayman
Cubic capacity/cylinder	2687 cc/6 cylinder
hp/kW	245/180
Construction period	from 2006
Number of produced cars	–

Volkswagen W12 Coupé

When the last variation of the concept car W12 was presented in 2001, nearly everything had basically remained the same. The W-12 engine that was fitted longitudinally in the direction of traffic in front of the rear axle still constantly drove all wheels. However, the cubic capacity of the machine had grown to six litres. This allowed the car to reach a maximum speed of 350 km/h – while W12 roadsters achieved "only" 290 km/h.

Model	Volkswagen W12 Coupé
Cubic capacity/cylinder	5998 cc/12 cylinder
hp/kW	600/441
Construction period	2001
Number of produced cars	unique piece

Volkswagen ConceptR

The design study ConceptR exemplifies what future sports cars made by VW could look like. Thus the designers chose an extravagant colour called "Frozen Sky" for its exterior and a leather interior coloured "Dark Earth" for the 416-centimetre long car as an eye-catcher – in contrast, the elements of the fascia are made of aluminium.

Model	Volkswagen ConceptR
Cubic capacity/cylinder	3189 cc/6 cylinder
hp/kW	265/195
Construction period	2003
Number of produced cars	unique piece

Volkswagen ConceptR

The ConceptR is powered by a V6 cylinder mid-engine and accelerates from 0 to 100 km/h in 5.3 seconds. Its top speed amounts theoretically to 270 km/h, however, is limited electronically to 250 km/h. By the way, as a special technical feature the engine sound can be controlled by a lever.

Model	Volkswagen ConceptR
Cubic capacity/cylinder	3189 cc/6 cylinder
hp/kW	265/195
Construction period	2003
Number of produced cars	unique piece

Volkswagen EcoRacer

In 2005, Volkswagen presented in Tokyo a prototype that could make history as the most fuel saving sports car ever. Its basic data are 3.4 litres of average fuel consumption and 230 km/h of maximum speed, its name is Eco-Racer. Eco because consumption and emissions are extremely low, and Racer because the dynamic potential of the mid-engine sports car is pure and fascinating.

Model	Volkswagen EcoRacer
Cubic capacity/cylinder	1484 cc/4 cylinder
hp/kW	136/100
Construction period	2004
Number of produced cars	unique piece

Volkswagen EcoRacer

The EcoRacer weighs only 850 kilograms thanks to its carbon fibre bodywork and accelerates in only 6.3 seconds to 100 km/h. The sports car powered by an newly developed turbo-diesel engine (!) offers a synthesis of uncompromising economy and sovereign sportiness. With a consciously fuel saving driving manner the car can travel more than 1000 kilometres on one tank filling (30 litres).

Model	Volkswagen EcoRacer
Cubic capacity/cylinder	1484 cc/4 cylinder
hp/kW	136/100
Construction period	2004
Number of produced cars	unique piece

Volkswagen Golf R 32

For the start of the fourth GTI generation (1998 until 2003), Volkswagen brought out its evergreen for the first time with a fully galvanised body. In 2001 – on the occasion of the 25th anniversary of the GTI – they launched a 180 hp turbo special model whose power output was topped in the same year by the Golf R32. At that time, the R32 was the strongest Golf ever developed with 241 hp.

Model	Volkswagen Golf R32
Cubic capacity/cylinder	3189 cc/6 cylinder
hp/kW	241/177
Construction period	2001 – 2003
Number of produced cars	–

Volkswagen Golf GTI

Numerous comfort lines make the GTI also in its fifth new edition (since 2004) the most popular model of its class. Under the bonnet of the current version sits a four-cylinder engine supported by turbo power which transfers its output via a six-step gear to the front wheels. The road performance is considerable – the maximum speed amounts to 233 km/h.

Model	Volkswagen Golf GTI
Cubic capacity/cylinder	1984 cc/4 cylinder
hp/kW	200/147
Construction period	from 2004
Number of produced cars	–

YES! Roadster

Funke & Will AG of Dresden which describes itself as a car manufacturer has been producing various individually made sports cars for some years. The first of these masterpieces, the YES! Roadster, was a fast gull wing door model with a maximum speed of 264 km/h with removable side windows. Like all YES! models, this car was also based on an aluminium space frame

Model	YES! Roadster
Cubic capacity/cylinder	1781 cc/4 cylinder
hp/kW	286/210
Construction period	2003 – 2006
Number of produced cars	–

YES! Roadster 3.2 Turbo

In comparison to the first release of Funke und Will AG, the YES! Roadster 3.2 Turbo has clearly grown bigger. Larger engines and a larger wheel base provide it with more overhang both in the front and the rear. This measure suits the pretty sports car exceptionally well. In combination with its wedge-shaped ascending side line, the 3.2 Turbo (281 km/h) appears substantially more stretched and more loaded with tension.

Model	YES! Roadster 3.2 Turbo
Cubic capacity/cylinder	3189 cc/6 cylinder
hp/kW	355/261
Construction period	from 2006
Number of produced cars	–

YES! Roadster 3.2

The YES! Roadster 3.2 is the perfect upgrade of the YES! that remains true to itself and its motto "Pure Driving Pleasure". The V6 engine under the bonnet of the two-seater provides the sports car that weighs just 640 kilogram with a top speed of 255 km/h and turns it into an all-out driving machine. The small series car is built, incidentally, in Großenhain near Dresden.

Model	YES! Roadster 3.2
Cubic capacity/cylinder	3189 cc/6 cylinder
hp/kW	255/188
Construction period	from 2006
Number of produced cars	–

Ascari KZ 1

Since 2005, a Dutch initiator and sports car enthusiast has been producing especially exclusive sports car, the Ascari KZ 1, in Great Britain. Only 50 units are produced manually, and buyers who would like to call one of them their own have to pay around EUR 350,000. The bodywork and chassis, incidentally, are made of Carbon, and it is powered by a V8 engine by BMW, while the car's maximum speed is approx. 320 km/h.

Model	Ascari KZ 1 (concentration camp 1)
Cubic capacity/cylinder	4941 cc/8 cylinder
hp/kW	500/368
Construction period	from 2005
Number of produced cars	max. 50

Aston Martin AMV 8 Concept

In January 2003 this prototype was shown at the North American International Auto Show at the fair display of Aston Martin. Almost two years later, the serial production of this design was started in an almost unchanged form (V8 Vantage). The show car is based, incidentally, on a new frame construction (aluminium platform) which can be used in a modified form for a great number of future models.

Model	Aston Martin AMV 8 Concept
Cubic capacity/cylinder	4280 cc/8 cylinder
hp/kW	385/283
Construction period	2002
Number of produced cars	unique piece

Aston Martin Twenty Twenty

The "Twenty Twenty" by automobile designer Giorgetto Giugiaro is a show car whose distinguishing characteristic is a partly visible aluminium frame. The concept car (446 centimetre length, 190 centimetre width and 130 centimetre height) is based on the Aston Martin DB 7 and also features a high-quality interior. Whether an Aston Martin will really look this way in 2020 remains to be seen.

Model	Aston Martin Twenty Twenty
Cubic capacity/cylinder	5935 cc/12 cylinder
hp/kW	500/368
Construction period	2000
Number of produced cars	unique piece

Aston Martin Zagato Vanquish Roadster

When the British noble brand Aston Martin co-operates with the Italian coach maker Zagato, one can be sure that the result will be especially exclusive and expensive. This collaboration also sometimes results in desirable unique pieces such as this Vanquish-based roadster with a top speed of 280 km/h.

Model	Aston Martin Zagato Vanquish Roadster
Cubic capacity/cylinder	5935 cc/12 cylinder
hp/kW	441/324
Construction period	2004
Number of produced cars	unique piece

Aston Martin Rapide

After a long break, Aston Martin decided to once again create a four-door, sportily inclined saloon. The result of this effort was called Rapide and was shown in 2006 at the Detroit Motor Show. Initially, the production of the concept car was not considered, however, due to the positive feedback the four-door model will soon be seen cruising the streets.

Model	Aston Martin Rapide
Cubic capacity/cylinder	5935 cc/12 cylinder
hp/kW	480/353
Construction period	2005
Number of produced cars	unique piece

Aston Martin Rapide

As a special highlight, the Rapide concept car features a big sunroof made of plastic. It allows the light to flood the generously laid-out interior of the four-door model, but can also be electrically darkened when required (!). The big sports coupé is started, incidentally, not by switching a key, but by pressing a start button made of glass!

Model	Aston Martin Rapide
Cubic capacity/cylinder	5935 cc/12 cylinder
hp/kW	480/353
Construction period	2005
Number of produced cars	unique piece

Aston Martin Vanquish

Aston Martin's current flagship, the twelve-cylinder Vanquish, became available in 2001. By the application of high-tech materials like aluminium and carbon the super car equipped with an electronic gas pedal tops the scales at only 1820 kilograms and reaches a maximum speed of 306 km/h. Its acceleration from 0 to 100 km/h requires only 4.7 seconds.

Model	Aston Martin Vanquish
Cubic capacity/cylinder	5935 cc/12 cylinder
hp/kW	450/330
Construction period	2001 – 2004
Number of produced cars	–

Aston Martin Vanquish S

In the Vanquish, the driver switches gears by means of a small paddle behind the steering wheel – just like in Formula 1 racers. Shifting gears within the transmission system takes only 250 milliseconds, which is less than the blink of an eye. The outer body of the Vanquish is made of aluminium. Every single metal sheet is adjusted manually – while each car is painted in the individual colour desired by the customer.

Model	Aston Martin Vanquish S
Cubic capacity/cylinder	5935 cc/12 cylinder
hp/kW	528/388
Construction period	from 2004
Number of produced cars	–

Aston Martin DB AR 1

The resemblance of the DB AR 1 Zagato to its closed counterpart, the Vantage Zagato, could not be denied – this vigorous roadster belonged to the DB 7 family. Initially, its career began as a concept car that was shown in January 2003 for the first time in Los Angeles. Given the positive feedback, Aston Martin resolved to produce this car with a top speed of 298 km/h in a small series.

Model	Aston Martin DB AR 1
Cubic capacity/cylinder	5935 cc/12 cylinder
hp/kW	440/324
Construction period	from 2003
Number of produced cars	40

Aston Martin DB 7 GT

In late 2002, the most powerful model of this series was released – the DB 7 GT, which received much attention. The optically very successfully updated car (maximum speed 298 km/h) could be ordered alternatively with a manual six-speed or an automatic five-speed gearbox. While the car moved in front on tyres measuring 245/40, the rear axle was equipped with tyres sized 265/35.

Model	Aston Martin DB 7 GT
Cubic capacity/cylinder	5935 cc/12 cylinder
hp/kW	435/320
Construction period	2003 – 2004
Number of produced cars	–

Aston Martin DB 7 Zagato

In 2002, a very special DB 7 model was developed with a slightly shortened Volante chassis. It was a very rare version whose looks were designed by the Italian coach maker Zagato. To enable more comfortable seating in the low riding coupé, the roof section received more headroom through the "double bubble" typical of Zagato car bodies.

Model	Aston Martin DB 7 Zagato
Cubic capacity/cylinder	5935 cc/12 cylinder
hp/kW	442/325
Construction period	2002 – 2003
Number of produced cars	99

Aston Martin DB 9 Coupé

The Aston Martin DB 9 not looks only hyper-modern – it is also produced in an equally modern and new factory in Gaydon, England. The body that appears monolithic allows the low riding 2+2 seater to appear fast even while standing still. And it is fast Indeed – 300 km/h are easily reached with its strong six-litre engine. According to its equipment, this fun car costs about EUR 155,000.

Model	Aston Martin DB 9 Coupé
Cubic capacity/cylinder	5935 cc/12 cylinder
hp/kW	457/336
Construction period	from 2004
Number of produced cars	–

Aston Martin DB 9 Volante

The Aston Martin brand, which has been part of the Ford Group since 1987, proves with the DB 9 Volante that high-calibre sports cars do not necessarily have to come from Italy. The body designers have refrained from using plastic parts in the DB 9 and instead relied only on the noblest materials – the leather interior finishing is available incidentally in 20 different colour hues.

Model	Aston Martin DB 9 Volante
Cubic capacity/cylinder	5935 cc/12 cylinder
hp/kW	457/336
Construction period	from 2004
Number of produced cars	–

Aston Martin V8 Vantage

The V8 Vantage is currently the "smallest" representative of the British noble brand. It exemplifies that size is not everything and can easily achieve a maximum speed of 280 km/h. Thanks to its ingenious weight distribution and good positioning of the centres of gravity (the gear is located near the rear axle) this high-capacity sports car can certainly keep up with its German and Italian competitors.

Model	Aston Martin V8 Vantage
Cubic capacity/cylinder	4280 cc/8 cylinder
hp/kW	385/283
Construction period	from 2005
Number of produced cars	–

Bentley Continental GT

The prestigious brand Bentley, which has been operating under the management of Volkswagen since 1998, presented the Bentley Continental GT in autumn 2002 – a coupé of the absolute top class. This first car developed under the aegis of VW presents a mixture of sporty dynamism and elegance. Like all Continental models, the GT features an interior that can be considered a masterpiece of craftsmanship.

Model	Bentley Continental GT
Cubic capacity/cylinder	5998 cc/12 cylinder
hp/kW	560/411
Construction period	from 2003
Number of produced cars	–

Bentley Continental GTC

The Continental GTC, the new 2+2 deluxe convertible of Bentley Motors, was introduced in April 2006. It completes the Continental series featuring the Continental GT (market launch in 2003) and the four-door model Continental Flying Spur (market launch in 2005). Although the Continental GTC reaches a maximum speed of 312 km/h, it will appeal primarily to customers who are looking for something else than a pure sports car.

Model	Bentley Continental GTC
Cubic capacity/cylinder	5998 cc/12 cylinder
hp/kW	560/411
Construction period	from 2006
Number of produced cars	–

Bentley Continental GTC

The Continental GTC convertible is powered by a mighty six-litre W12 engine with a dual turbo charger which already disposes of plenty of torque at a mere 1600 rotations per minute. The Continental GTC is equipped with a traction control system and an electronic stabilisation program (ESP) – the driver can switch off both, if requested.

Model	Bentley Continental GTC
Cubic capacity/cylinder	5998 cc/12 cylinder
hp/kW	560/411
Construction period	from 2006
Number of produced cars	–

Bristol Blenheim 3

In the course of model development, the Blenheim, which was already launched in 1993 was twice slightly revised optically and technically before the third edition was released in 2002. The simple, but very sporty coupé with just 290 centimetres of wheel base is very luxuriously equipped inside and reaches a maximum speed of 250 km/h.

Model	Bristol Blenheim 3
Cubic capacity/cylinder	5898 cc/8 cylinder
hp/kW	no information
Construction period	from 2002
Number of produced cars	–

Bristol Blenheim Speedster

Based on the Blenheim coupé (290 centimetres of wheel base) Bristol created for convertible fans in search of something really unusual the Roadster with a top speed of 257 km/h. The design of the front part of the two-seater vehicle (absolute height only 112 centimetres) is reminiscent of the Type 405 built in the 1950s.

Model	Bristol Blenheim Speedster
Cubic capacity/cylinder	5898 cc/8 cylinder
hp/kW	no information
Construction period	from 2003
Number of produced cars	–

Bristol Fighter

In 2002, the Bristol Fighter was still a design study. Underneath its light aluminium body with upwards swinging gull-wing doors can be found the technology of the Dodge Viper. Two years later when the Fighter with a top speed of 310 km/h went into production, nothing had changed in the combination of British understatement and American engine technology.

Model	Bristol Fighter
Cubic capacity/cylinder	7996 cc/10 cylinder
hp/kW	558/410
Construction period	from 2004
Number of produced cars	–

Bristol Fighter S

Among aeroplane fans, the name "Fighter" certainly provides great enthusiasm, because in 1918 Bristol sent the first Fighter up in the air. Since 2004, Bristol has been building a sports car under this model name that undoubtedly belongs to the class of élite vehicles. The price also matches its exclusivity and rarity – the car is available at a starting price of EUR 315,000.

Model	Bristol Fighter S
Cubic capacity/cylinder	7996 cc/10 cylinder
hp/kW	669/558
Construction period	from 2004
Number of produced cars	–

Caparo T1

The car that looks obviously like a Formula 1 racer, is nothing else than a two-seater prototype by the car manufacturer Caparo. In the opinion of the British manufacturer, this is what the streetcars of the future will look like! Weighing only 500 kilograms, the light Caparo T1 was shown for the first time to the general public in May 2006 at the Top Marques Luxury Fair in Monaco and is planned to go into production in late 2006.

Model	Caparo T1
Cubic capacity/cylinder	2400 cc/8 cylinder
hp/kW	476/350
Construction period from	2006
Number of produced cars	–

Caparo T1

Materials used in aeroplane and racer construction help the Caparo T1 to achieve power-to-weight ratio that is double that of the 1001 hp Bugatti Veyron. The T1 accelerates in less than three seconds to 100 km/h. In normal everyday use – i.e. on the street – even with a breezy driving manner a tank filling (70 litres) should last for about 800 kilometres.

Model	Caparo T1
Cubic capacity/cylinder	2400 cc/8 cylinder
hp/kW	476/350
Construction period	from 2006
Number of produced cars	–

Caterham Seven CSR

Headquartered in Dartford near London, U.K., Caterham Cars Ltd has been further developing the former Lotus Seven. Since 1990, the official Seven successor was optimised several times and technically refined. In addition to the models with 106 hp and 122 hp the class was complemented in 2004 by the even stronger CSR models (203 hp and 264 hp) – that can achieve speeds up to 250 km/h.

Model	Caterham Seven CSR
Cubic capacity/cylinder	2261 cc/4 cylinder
hp/kW	264/194
Construction period	from 2004
Number of produced cars	

Farboud GTS

Arash Farboud founded in 2004 in the British Wiltshire County a small, but excellent automobile manufacture company that produces a sports car named after him. To control the production costs, the Farboud GTS is normally equipped with a 275 hp Ford V6 engine. Alternatively there is also a "supercharged" version available featuring 375 hp.

Model	Farboud GTS
Cubic capacity/cylinder	2968 cc/6 cylinder
hp/kW	275/202
Construction period	from 2005
Number of produced cars	–

Invicta S1-320

The sports car made by the Invicta brand already existed in the 1930s. They were definitely exotic and accordingly expensive and rare. The in the meantime reactivated brand would like to continue this tradition – the prototype shown for the first time in 2002 has become a serial model that is available in three output levels starting at EUR 110,000.

Model	Invicta S1-320
Cubic capacity/cylinder	4601 cc/8 cylinder
hp/kW	325/239
Construction period	from 2002
Number of produced cars	–

Invicta S1-600

Even in its basic version (S1-320) the British Invicta is already a rather fast vehicle with a top speed of 273 km/h. The next stronger variation (S1-420) is powered by 427 hp. Currently, the S1-600 reigns as the head of the model range. It also uses the beefy Ford engine, which in this case, however, is even more powerful with the help of a compressor – its strength allows the car to reach 329 km/h.

Model	Invicta S1-600
Cubic capacity/cylinder	4601 cc/8 cylinder
hp/kW	612/450
Construction period	from 2005
Number of produced cars	–

Jaguar R Coupé Concept

Among the attractions of the Jaguar display at the Frankfurt International Motor Show of 2001 this bronze-metallic painted coupé was greatly admired. Its shape originated on the drawing board of Ian Callum, Jaguar's head of design. It is a pity that this study will not go into production – after all, a silver-plated radiator grille adorns its front, and the rear even features two silver-plated tank lids.

Model	Jaguar R Coupé Concept
Cubic capacity/cylinder	4196 cc/8 cylinder
hp/kW	400/294
Construction period	2000
Number of produced cars	unique piece

Jaguar XJR Alu-Showcar

Since the debut of the first XJ 6 in September 1968, about 800,000 vehicles of the XJ series were produced until 2003. The seventh generation offered since 2003 featured for the first time an aluminium monocoque body. Although the car became simultaneously longer, higher and wider, its weight could be reduced by about 200 kilograms compared to a steel construction.

Model	Jaguar XJR Alu-Showcar
Cubic capacity/cylinder	4196 cc/8 cylinder
hp/kW	298/219
Construction period	2002
Number of produced cars	unique piece

Jaguar XF 10

In 2003, the Spanish design studio Fuore Design presented at the International Motor Show in Barcelona the XF 10, a slightly unusual Jaguar interpretation. The already originally sporty car was upgraded with an even more fascinating design with unmistakable sportiness. Derived from the Formula 1, it is the embodiment of pure strength and dynamism.

Model	Jaguar XF 10
Cubic capacity/cylinder	7000 cc/10 cylinder
hp/kW	640/471
Construction period	2003
Number of produced cars	unique piece

Jaguar BlackJag

Fuore design continued the philosophy of the XF 10 with its successor, the BlackJag. While the XF 10 was an expression of pure sportiness, the BlackJag had a rather more dignified appearance. According to the company press release, it consists of a sports car that embodies the traditional elegance of the British automotive brand and combines it with their engagement in the Formula 1.

Model	Jaguar BlackJag
Cubic capacity/cylinder	7000 cc/10 cylinder
hp/kW	640/471
Construction period	2004
Number of produced cars	unique piece

Jaguar BlackJag

Compared to the XF 10, the exterior of the BlackJag was generally redesigned and rounded to provide this concept car with an independent elegance. For example, the air inlet vents fit more fluently into the whole silhouette of the car and are integrated more harmoniously in its whole structure. Fuore design literally stated: "The name BlackJag suggests elegance, which has been the driving element of our design study".

Model	Jaguar BlackJag
Cubic capacity/cylinder	7000 cc/10 cylinder
hp/kW	640/471
Construction period	2004
Number of produced cars	unique piece

Jaguar Advanced Lightweight Coupé

The 2 + 2 concept car developed by the Jaguar Advanced Design Team is the expression of their vision of what the next generation of Jaguar models could look like – extremely beautiful with fluently drawn lines in combination with athletic elements. This study presents an absolutely new definition of the coupé design – in front as well as the rear there is no body overhang, which provides the car with an especially powerful appearance.

Model	Jaguar Advanced Lightweight Coupé
Cubic capacity/cylinder	n.s./8 cylinder
hp/kW	no information
Construction period	2004
Number of produced cars	unique piece

Jaguar Advanced Lightweight Coupé

In 2005 at the North American International Motor Show, Jaguar introduced the Advanced Lightweight Coupé, which is undoubtedly one of the most interesting sports car studies of the past few years. Innovative technologies (aluminium bodywork) in combination with high-quality design details (21-inch aluminium rims) distinguish this unique vehicle that can theoretically reach a speed of 290 km/h.

Model	Jaguar Advanced Lightweight Coupé
Cubic capacity/cylinder	n.s./8 cylinder
hp/kW	no information
Construction period	2004
Number of produced cars	unique piece

Jaguar XJR 4.2 V8 Compressor

For many loyal Jaguar fans, the XJ saloon remains the ideal means of locomotion. In the current model range the model name XJ 6 is experiencing a comeback with the 3-litre V6 engine. Other engines include a 3.5-litre V8, a 4.2-litre V8 and a 4.2-litre V8 compressor engine with 395 hp – the latter model is used in the prestigious segment of sports saloons.

Model	Jaguar XJR 4.2 V8 Compressor
Cubic capacity/cylinder	4196 cc/8 cylinder
hp/kW	298/219
Construction period	from 2004
Number of produced cars	–

Jaguar XK 8

Jaguar sports cars have always been "advanced". In 2005, (after a break of more than 40 years) the company reintroduced the production of lightweight sports cars under the title "lightweight". As opposed to other cars, a Jaguar is not based on a space frame construction. Its body is put together as a unibody construction of riveted and bonded metal sheets.

Model	Jaguar XK 8
Cubic capacity/cylinder	4196 cc/8 cylinder
hp/kW	300/224
Construction period	from 2006
Number of produced cars	–

Jaguar XKR Silverstone

To the great joy of drivers with sporty ambitions, Jaguar presented at the Geneva Motor Show in 1998 the XK for the first time as a compressor version. This model called XKR was distinguished by two slit bonnet louvers. In 2000, the XKR Silverstone was launched, a special series limited to 200 vehicles that were exclusively coloured "Platinum".

Model	Jaguar XKR Silverstone
Cubic capacity/cylinder	3996 cc/8 cylinder
hp/kW	363/267
Construction period	2000
Number of produced cars	200

Jaguar XKR 100

In September 2001, Sir William Lyons, the founder of Jaguar would have turned 100. On the occasion of this significant birthday, the company launched the XKR 100 on the market, a special model equipped with a compressor engine. The coupé with a maximum speed of 250 km/h was available in Germany at that time for DM 195.000, while the convertible version cost DM 205.000.

Model	Jaguar XKR 100
Cubic capacity/cylinder	3996 cc/8 cylinder
hp/kW	363/267
Construction period	2001
Number of produced cars	500

Jaguar XKR Coupé

With the XKR presented in summer 2006, Jaguar complemented the successful XK class with another model. It is available as a coupé (starting at EUR 94.990) and also as a convertible (starting at EUR 102.990) and is equipped with a compressor engine. The six step automatic gear is operated by modern rocker switches and enables the XKR to accelerate from 0 to 100 km/h in 5.2 seconds.

Model	Jaguar XKR
Cubic capacity/cylinder	4196 cc/8 cylinder
hp/kW	416/306
Construction period	from 2007
Number of produced cars	–

Jaguar XKR Cabriolet

The XKR underlines its sporty purpose by a different front fender, an aluminium finish for the upper wire mesh grille, lateral air intake vents with an aluminium look, bonnet louvers as well as black brake callipers featuring an "R" logo. In addition, the XKR is especially sportive with a four pipe exhaust system.

Model	Jaguar XKR
Cubic capacity/cylinder	4196 cc/8 cylinder
hp/kW	416/306
Construction period	from 2007
Number of produced cars	–

Lotus Elise Sport Racer

Limited special series always are a special attraction to car collectors. The distinguishing features of this model include above all its remarkable paint in Arden red with white stripes as well as black leather seat covers. A newly co-ordinated chassis, a new pedal system and especially light aluminium rims complete the car's equipment.

Model	Lotus Elise Sport Racer
Cubic capacity/cylinder	1796 cc/4 cylinder
hp/kW	192/141
Construction period	2005
Number of produced cars	199

Lotus Exige

Most simply stated, the Exige is the coupé version of the successful Lotus Elise. It is also based on a light alloy chassis, and the engine is transversely installed in front of the rear axle. In the basic version, the 380-centimetres compact fun mobile is powered by 192 hp, while the slightly more vicious compressor variation features 220 hp.

Model	Lotus Exige
Cubic capacity/cylinder	1796 cc/4 cylinder
hp/kW	192/141
Construction period	from 2005
Number of produced cars	–

Lotus Europa S

Lotus already featured a model called "Europa" in its program from 1967 to 1975. Now they are continuing the tradition of the series of that time, because the new Europa with a maximum speed of 225 km/h also consists of a low and closed coupé with high-quality interior furnishing (all in leather).

Model	Lotus Europe S
Cubic capacity/cylinder	2000 cc/4 cylinder
hp/kW	203/149
Construction period	from 2006
Number of produced cars	–

Marcos TSO Convertible

Those who are declared dead live longer – this proverb definitely applies to the sports car manufacturer Marco that has already been reactivated several times. In 1964, the British company introduced its first successful model, the 1800 GT. As of 1973 the production was ceased before a new attempt was made in the early 1980s. Since 2004, the company is trying to make another comeback with the TSO model.

Model	Marcos TSO Convertible
Cubic capacity/cylinder	5665 cc/8 cylinder
hp/kW	355/260
Construction period	from 2004
Number of produced cars	–

Marcos TSO GT2

In terms of its technology, the Marcos GT2 is nothing more than a closed variation of the TSO Roadster. Both models (with a wheel base of 228 centimetres) are equipped with a Chevrolet engine which is available in the output levels 355 hp, 405 hp and 480 hp. Accordingly, the maximum speed varies between 275 km/h and 300 km/h.

Model	Marcos TSO GT2
Cubic capacity/cylinder	5665 cc/8 cylinder
hp/kW	480/353
Construction period	from 2005
Number of produced cars	–

MB Roadcar

The English company MB Roadcars is one of the many new sports car brands that were only founded after the turn of the millennium. For a change, MB constructed at one point constructed an extremely low-riding racer in which three persons can sit comfortably side by side in front. The V8 engine of the road car is very powerful, incidentally – the red section begins only at 11.000 rpm.

Model	MB Roadcar
Cubic capacity/cylinder	3995 cc/8 cylinder
hp/kW	500/368
Construction period	from 2007
Number of produced cars	–

MG TF

With the insolvency of MG Rover, the production of the MG TF which it had launched in the meantime was discontinued in April 2005. The rights to the MG Design went already in late 2004 to a Chinese group of companies. Supposedly in Spring 2007 the Chinese want to once again build a MG Roadster in a slightly modified form – the production site will most probably be once again Longbridge in England.

Model	MG TF
Cubic capacity/cylinder	1796 cc/4 cylinder
hp/kW	160/118
Construction period	2002 – 2005
Number of produced cars	approx. 44.000

MG SV-R

To position a potent image bearer on the market, MG Rover developed a sports coupé that cost approximately EUR 130,000 – the SV-R. Its bodywork is made of carbon and is "agglomerated" from about 5000 individual parts. Under the bonnet, a Ford Visteon engine from light alloy is at work – it provides the SV-R with a maximum speed of 283 km/h.

Model	MG SV-R
Cubic capacity/cylinder	4995 cc/8 cylinder
hp/kW	385/283
Construction period	2004 – 2005
Number of produced cars	–

MG SV-R

As a pedigree sports device, the MG SV-R is equipped rather sparingly – since every gram would impact its agility too much. Therefore, one takes a seat on the narrow bucket seats whose only luxury is their cover of leather and Alcantara. Due to the bankruptcy of the MG Rover Group, this two-seater was already discontinued in spring 2005.

Model	MG SV-R
Cubic capacity/cylinder	4995 cc/8 cylinder
hp/kW	385/283
Construction period	2004 – 2005
Number of produced cars	–

Mini Cooper S Cabrio

With the Mini Cooper S Cabrio, the manufacturer launched in 2004 the to date strongest version of the popular small Briton. The compact power-pack is especially sporty and powered by a four-cylinder compressor engine with intercooling. It is serially equipped with a six-step transmission. The maximum speed of the open Mini is 215 km/h.

Model	Mini Cooper S Cabrio
Cubic capacity/cylinder	1598 cc/4 cylinder
hp/kW	170/125
Construction period	from 2004
Number of produced cars	–

Mini Cooper S John Cooper

Under the management of BMW, Mini has by now launched eight different models on the market. In Spring 2006, they launched an especially collectible variation, the Mini Cooper S with John Cooper Works GP Kit. This limited edition sporty little streak of lightning (acceleration from 0 to 100 km/h in 6.5 seconds) is offered exclusively in the colour combination "Metallic Thunder Blue" and "Pure Silver Metallic" (roof).

Model	Mini Cooper S John Cooper
Cubic capacity/cylinder	1598 cc/4 cylinder
hp/kW	218/160
Construction period	2006
Number of produced cars	2000

Mini Concept Geneva

The Mini "Concept Geneva" was launched on the occasion of the [illegible] birthday of the creator of the Mini, Sir Alec Issigonis in the year 2006 and to commemorate the victories of the Mini brand at the Monte Carlo Rally in the mid 1960s. With this interpretation, the model versions Traveller, Countryman and Clubman Estate were shown as trend-setting new model variations.

Model	Mini Concept Geneva
Cubic capacity/cylinder	1598 cc/4 cylinder
hp/kW	218/160
Construction period	2006
Number of produced cars	unique piece

Morgan Aero 8

The Morgan Aero 8, introduced at the Geneva Salon in 2000, has nothing in common with its namesake built from 1919 to 1936, because a V2 engine powered the early Aero. The perky newcomer with a top speed of 258 km/h moves on 18-inch magnesium wheels and has a modern BMW V8 engine hidden under its bonnet– it accelerates the Aero 8 after 4.9 seconds to 100 km/h.

Model	Morgan Aero 8
Cubic capacity/cylinder	4398 cc/8 cylinder
hp/kW	286/210
Construction period	2001 – 2004
Number of produced cars	296

Morgan Aero 8

As part of the model development process, all Aero 8 cars made from 2004 received some cosmetic changes. Buyers of this version profited from a slightly enlarged interior as well as more boot volume. To comply with US certification requirements, it additionally required some of engine engineering modifications.

Model	Morgan Aero 8
Cubic capacity/cylinder	4398 cc/8 cylinder
hp/kW	330/243
Construction period	2004 – 2006
Number of produced cars	approx. 120

Morgan Aero 8

Whether the Aero 8 had really needed a face-lift for the model year in 2006 is purely a matter of opinion. The renewed modification of its appearance was mainly limited to the front part (extended radiator grille, new lamps) as well as the rear area. Under the aluminium body, everything remained as before – the Roadster with the retro look continues to be powered by a BMW-V8 engine.

Model	Morgan Aero 8
Cubic capacity/cylinder	4398 cc/8 cylinder
hp/kW	330/243
Construction period	from 2006
Number of produced cars	–

Morgan V6

Up to the end of production in 2004, Morgan was able to sell its classic "+8" model about 6000 times. The successor who has been leaving the factory halls in Malvern Link since Summer 2006 carries the name Morgan V6 and continues to be equipped with the body of the "+8". However, today a six-cylinder engine by Ford is at work under the long hood instead of the former eight-cylinder engine.

Model	Morgan V6
Cubic capacity/cylinder	2967 cc/6 cylinder
hp/kW	226/166
Construction period	from 2004
Number of produced cars	

Morgan Aeromax Concept

On the Geneva Motor Show in 2005, the British sports car manufacturer Morgan presented the "Aeromax", a vehicle that was simultaneously a concept car and remittance work, because the unique piece with a top speed of 260 km/h was primarily made for a Swiss banker. The shape of the extravagant coupé – strictly following the style of the 1930s – is, incidentally, a draft of the design student Matt Humphries.

Model	Morgan Aeromax Concept
Cubic capacity/cylinder	1398 cc/8 cylinder
hp/kW	330/243
Construction period	2004
Number of produced cars	unique piece

Morgan 70th

The vehicles of the special series "70th" which Morgan introduced on the occasion of the 70th anniversary of the 4/4 will look rather different from each other for a special reason: their look is adapted to the respective decade that they represent. These details appear above all in the design of the front part, the car's colour, and of course in the interior finishing.

Model	Morgan 70th
Cubic capacity/cylinder	1796 cc/4 cylinder
hp/kW	114/84
Construction period	2006
Number of produced cars	142

Morgan 70th

In 2006 Morgan celebrated a very special anniversary: its four-cylinder model – the Morgan 4/4 – turned 70! For this reason, the British sports car manufacturers issued a limited special series in honour of the longest continuous production of a car model. The small series is constructed in such a way that two vehicles represent each year.

Model	Morgan 70th
Cubic capacity/cylinder	1796 cc/4 cylinder
hp/kW	114/84
Construction period	2006
Number of produced cars	142

Morgan +4

The current Morgan +4 still carries on the tradition of the famous class winner of Le Mans (1962). It has remained loyal to the classical way of manufacturing and is based on a galvanised chassis, while the body construction is produced from a combination of ash wood and aluminium sheet. Weighing only 880 kilograms, the roadster has a maximum speed of 180 km/h.

Model	Morgan +4
Cubic capacity/cylinder	1796 cc/4 cylinder
hp/kW	125/92
Construction period	from 2006
Number of produced cars	–

Morgan +4 four-seater

For decades Morgan has been offering in addition to its two-seater roadsters also a slightly more spacious variation, the four-seater. This niche model with its proportions that need some getting used to (long wheel base) was rather hard to find. However, this changed in 2006. At that point, the four-seater that costs about EUR 60,000 is powered by a V6 engine and operates at a maximum speed of 215 km/h.

Model	Morgan +4 four-seater
Cubic capacity/cylinder	2967 cc/6 cylinder
hp/kW	226/166
Construction period	from 2006
Number of produced cars	–

Noble M 12 GTC

Lee Noble, a racing driver by profession, as well as racing expert and inventor started his own business in 1985. He initially worked on optimising various sports cars. In the back of his mind he already had the idea of creating his own car – this dream became a reality in 1999 with the M 10. One year later, the first car produced in a small series, the Noble (M 12) enriched the sports car scene.

Model	Noble M 12 GTC
Cubic capacity/cylinder	2968 cc/6 cylinder
hp/kW	290/216
Construction period	from 2000
Number of produced cars	–

Noble M 12 GTO 3R

Nowadays the trade name Noble has become a fixed concept among sports car fans – the British manufacturers have applied all possible means to make their sports cars internationally known. According to factory specifications, the aim was to build about 200 cars annually and then to stop. Under the bonnet of the Noble, incidentally, a Ford engine is at work.

Model	Noble M 12 GTO 3R
Cubic capacity/cylinder	2968 cc/6 cylinder
hp/kW	352/262
Construction period	from 2001
Number of produced cars	–

Noble M 14

Compared to earlier models, the space frame chassis of the M 14 has become more rigid. This was necessary to be able to fulfil the wish of many customers for even more performance. The engine power of the M 14 with a maximum speed of approximately 280 km/h is transmitted via a six-step gear to the wide back wheels – while acceleration from 0 to 100 km/h takes 4 seconds.

Model	Noble M 14
Cubic capacity/cylinder	2968 cc/6 cylinder
hp/kW	400/298
Construction period	from 2004
Number of produced cars	–

Noble M 15

With the M 15 Noble a car was introduced to the market which could not only be used in racing, but also as a touring vehicle. Therefore, a satellite navigation system belongs to its basic equipment, and it features is a boot which can accommodate more than only the spare tyre. The speedometer needle of the M 15 stops just short of 300 km/h, however, this will certainly change soon.

Model	Noble M 15
Cubic capacity/cylinder	2968 cc/6 cylinder
hp/kW	455/335
Construction period	from 2006
Number of produced cars	–

Noble M 400

Currently, the financial entrance level to the category of noble cars is at around EUR [illegible] 000. For this amount one can purchase the M 400. Like the other models of the manufacturer, it has been conceived as a 2+2-seater coupé. However due to its slightly more elevated body (114 centimetres) it does not look quite as sleek as an M15. Under the bonnet of the M 400, a V6 turbo engine by Ford is at work.

Model	Noble M 400
Cubic capacity/cylinder	2968 cc/6 cylinder
hp/kW	425/313
Construction period	from 2006
Number of produced cars	–

TVR Tuscan

The sports model by TVR, the Tuscan with a top speed of more than 255 km/h, was shown to the general public already in 1996 in its prototype stage – while its serial production started four years later. The light plastic body rests on a solid tubular steel frame with a wheel base of 236 centimetres. The Tuscan is offered in several performance levels with a 3.6-litre and 4.0-litre engine.

Model	TVR Tuscan
Cubic capacity/cylinder	3605 cc/6 cylinder
hp/kW	355/261
Construction period	from 2000
Number of produced cars	–

TVR Sagaris

While some TVR fans were shocked at first at the appearance of the new Sagaris (top speed in excess of 295 km/h), others found its aggressive styling immediately fascinating. After all, the hood of this super sports vehicle is literally "dissected" by 17 air intakes. The reason for this is the crude six-cylinder engine, which develops not only a great amount of strength but also a great amount of heat.

Model	TVR Sagaris
Cubic capacity/cylinder	3996 cc/6 cylinder
hp/kW	385/283
Construction period	from 2005
Number of produced cars	–

Virago

The British company Virago Cars that was founded only in 2005, aimed to enrich the sports car market with a two-seater coupé. The production preparations for the Virago processed rapidly– when the first cars left the factory at the end of 2006, 16 select retailers on the British isle handled the distribution.

Model	Virago
Cubic capacity/cylinder	3000 cc/6 cylinder
hp/kW	230/169
Construction period	from 2006
Number of produced cars	–

Virago

According to its factory specifications, the Virago that is available in two engine versions is a quite normal everyday car, which offers a lot of driving pleasure. The low-riding coupé is extensively equipped serially (navigation system, air-conditioning) and is painted only in metallic hues. The price of the car with a top speed of 240 km/h amounts to about 50.000 British pounds.

Model	Virago
Cubic capacity/cylinder	3000 cc/6 cylinder
hp/kW	280/206
Construction period	from 2006
Number of produced cars	–

Bugatti EB 16.4 Veyron

Since December 2000, Bugatti has been a member of the Volkswagen group. On the occasion of this re-establishment, Bugatti Automobiles S.A.S. showed at the 71st Geneva International Motor Show for the first time the sports cars EB 16.4 Veyron. For this 1001 hp model, the future was planned to begin in 2003, however, for technical reasons the launch was moved to autumn 2005.

Model	Bugatti EB 16.4 Veyron
Cubic capacity/cylinder	7993 cc/16 cylinder
hp/kW	1001/736
Construction period	from 2005
Number of produced cars	–

Peugeot 907

The 907 concept car introduced at the Paris Motor Show in 2004 was supposed to provide a model of a possible future Peugeot sports car. The two-seater coupé is powered by a six-litre V12 engine. The power of this longitudinally positioned front mid-engine is transmitted to the wheels via a six-step gearbox placed before the rear axle.

Model	Peugeot 907
Cubic capacity/cylinder	6000 cc/12 cylinder
hp/kW	500/368
Construction period	2004
Number of produced cars	unique piece

Bugatti EB 16.4 Veyron

According to factory specifications, the Veyron 16.4 is the world's most innovative high-capacity sports car. The sports coupé with a high speed of more than 400 km/h is powered by a 16-cylinder mid-engine with four turbochargers. The engine has an output of 1001 hp – the immense power is portioned out by the "world's fastest switch gear unit", a special direct switch gearbox with seven forward gears.

Model	Bugatti EB 16.4 Veyron
Cubic capacity/cylinder	7993 cc/16 cylinder
hp/kW	1001/736
Construction period	from 2005
Number of produced cars	–

Peugeot 907

The Peugeot 907 coupé has a wheel base of 250 centimetres. All in all, the elegant car is 437 centimetres long, 188 centimetres wide and a modest 121 centimetres high. The passenger's cabin that fits two persons, was deliberately shifted far to the back to create space for the long bonnet – the weight of the ready-to-drive car is 1400 kilograms.

Model	Peugeot 907
Cubic capacity/cylinder	6000 cc/12 cylinder
hp/kW	500/368
Construction period	2004
Number of produced cars	unique piece

Renault Clio V6

In 2003, the second generation of the Clio Sport was ready to be launched. While the name was reduced to Renault Clio V6, the car's performance and maximum speed (now 250 km/h) were increased. New fenders and a track widened in the front gave the Clio V6 an even more dynamic character – its wide 18-inch wheels are already included serially.

Model	Renault Clio V6
Cubic capacity/cylinder	2946 cc/6 cylinder
hp/kW	254/187
Construction period	2003 – 2005
Number of produced cars	1288

Renault Clio Sport 2.0 16 V

Since 2006, the new Clio Sport 2.0 16 V continues the tradition of its predecessors. The powerfully designed car offers a wealth of technology highlights, in particular, the "diffuser" integrated into the rear fender. This air baffle element accelerates the airflow under the car, which creates a vacuum that sucks the vehicle to the road.

Model	Renault Clio Sport 2.0 16 V
Cubic capacity/cylinder	1998 cc/4 cylinder
hp/kW	197/145
Construction period	from 2006
Number of produced cars	–

Renault Mégane Sport F1

In honour of world champion Fernando Alonso, Renault in co-operation with the magazine "sport auto" introduced the special model "Mégane Sport F1 Team sport auto edition". The attractive car optically resembles the design of the successful Renault F1 team and was conceived for ambitious drivers who place great value on high curve speeds.

Model	Renault Mégane Sport F1
Cubic capacity/cylinder	1998 cc/4 cylinder
hp/kW	224/165
Construction period	2006
Number of produced cars	100

Renault Mégane Sport

With a dynamic look and many comfort details, the Mégane Sport has initiated its next life cycle. The latest generation of the top model of the Mégane class can be recognised by its headlights, which combine with the V-shaped radiator grille to form a unit. Rear lights with a transparent look dominate the rear view.

Model	Renault Mégane Sport
Cubic capacity/cylinder	1998 cc/4 cylinder
hp/kW	224/165
Construction period	from 2006
Number of produced cars	–

Renault Talisman

With the Talisman, a four-seater sport coupé concept car, Renault presented the design of a high-class model at the 59th International Motor Show in Frankfurt. The 4.80 metre long Talisman offers the latest in technology and high-calibre electronics in every aspect. For example, the interior is accessed by two 2.45 metre-wide gull-wing doors that swing upward in an electrohydraulic fashion – controlled by an electronic chip card.

Model	Renault Talisman
Cubic capacity/cylinder	4494 cc/8 cylinder
hp/kW	340/250
Construction period	2001
Number of produced cars	unique piece

Renault Talisman

The interior of the Talisman offers a surprising combination of understated elegance coupled with highly complex high-tech equipment. This includes bucket seats with a substructure made of carbon fibre and red leather covers. The seat height can be infinitely variably adjusted with the help of integrated aerial cushions, which guarantees optimum seating comfort.

Model	Renault Talisman
Cubic capacity/cylinder	4494 cc/8 cylinder
hp/kW	340/250
Construction period	2001
Number of produced cars	unique piece

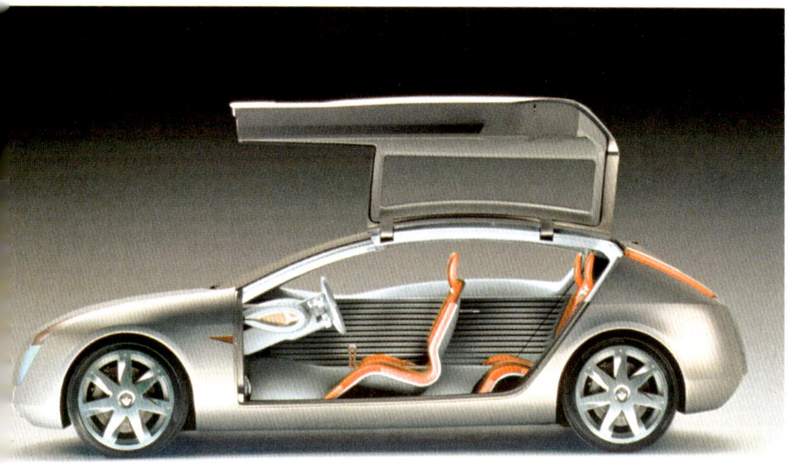

Renault Wind

Compact measurements and the innovative 2+1-seating concept distinguish the "Wind" roadster concept car with which Renault gained a large amount of attention in 2004. The only 3.87 metre long sports car has a fascinating appeal with its combination of razor sharp lines and flowing curves. Decorative elements of the purist body design include almond-shaped rear lights and headlights that are positioned deep within the fenders.

Model	Renault Wind
Cubic capacity/cylinder	1998 cc/4 cylinder
hp/kW	135/99
Construction period	2003
Number of produced cars	unique piece

Renault Fluence

The Renault concept car Fluence redefines the concept of the luxurious four-seater sports coupé. Sensuous forms dominate the fluent lines and friendly interior of this car. To Renault design manager Patrick le Quément, the Fluence is among the heirs of the long tradition of the French art of coach making. Renault literally states that the "Fluence is equally a drawing and a sculpture".

Model	Renault Fluence
Cubic capacity/cylinder	3499 cc/6 cylinder
hp/kW	280/204
Construction period	2004
Number of produced cars	unique piece

Renault Altica

With the Altica estate car, Renault has created a concept car whose dynamic appearance is distinguished by a long hood and distinctive curves. The Altica connects the utility value of an estate car with the sporty charm of a coupé and high travel comfort. In short: The vehicle embodies the independent vision of the estate car of the future.

Model	Renault Altica
Cubic capacity/cylinder	1995 cc/4 cylinder
hp/kW	177/130
Construction period	2006
Number of produced cars	unique piece

Venturi Fetish

In summer 1997, the Venturi production sites shut down for the first time. While a Thai investor managed to get the enterprise going again, nevertheless in the year 2000 it was shut down. The Venturi S.A. was acquired afterwards by Gildo Pallanca Pastor. He moved the company headquarters to Monaco, continued the spare part delivery service and developed a new model, the Fetish.

Model	Venturi Fetish
Cubic capacity/cylinder	electric motor
hp/kW	245/180
Construction period	from 2004
Number of produced cars	–

Venturi Fetish

The Fetish prototype shown in 2002 was moved by a four-cylinder engine. To the surprise of automobile enthusiasts, the design of the small series was altered once more and equipped as of 2004 with an electric motor. The light sports car with an aluminium chassis and light alloy body requires 4.5 seconds to accelerate from 0 to 100 km/h and reaches a maximum speed of 170 km/h.

Model	Venturi Fetish
Cubic capacity/cylinder	electric motor
hp/kW	245/180
Construction period	from 2004
Number of produced cars	–

Alfa Romeo Spider

At the Geneva International Motor Show of 1995, Alfa Romeo showed a Spider with an absolutely new styling. Together with its sister model, the Alfa Romeo GTV (coupé), it continued the successful co-operation between the Centro Stile Alfa Romeo and Pininfarina. Once again, with this model the co-operation resulted in an exemplary design interpretation of the timeless sports car dynamism special to the Alfa Romeo brand.

Model	Alfa Romeo Spider
Cubic capacity/cylinder	1970 cc/4 cylinder
hp/kW	150/110
Construction period	1995 – 1998
Number of produced cars	–

Alfa Romeo Spider 3.2 V6

Within the scope of model development, the Alfa Romeo Spider presented itself as of 2003 in a perfected outfit. In addition, the sports car was offered for the first time with the high-tech direct fuel injection 2.0 JTS that complies with the strict exhaust emission standard EURO-4. However, the fascination of this model that has always embodied a symbiosis of innovation and tradition remained unchanged.

Model	Alfa Romeo Spider 3.2 V6
Cubic capacity/cylinder	3179 cc/6 cylinder
hp/kW	240/177
Construction period	2003 – 2004
Number of produced cars	–

Alfa Romeo Spider

In summer 2003, the entrance-level price of the classical Alfa Romeo Spider was around EUR 26,800. The model was moderately retouched under the direction of Pininfarina and received a slightly enlarged unique radiator grille with noble chrome-plated crossbeams and featuring a plaque high on top with the Alfa Romeo brand name.

Model	Alfa Romeo Spider
Cubic capacity/cylinder	1970 cc/4 cylinder
hp/kW	166/122
Construction period	2003 – 2004
Number of produced cars	–

Alfa Romeo Spider

In summer 2006, Alfa Romeo continued the tradition of open sports cars with its newly contrived convertible model. The Spider is available as a 2.2-litre model (185 hp) as well as a 3.2-litre variation. The latter disposes of a power output of 260 hp and is equipped with a permanent four-wheel drive. Depending on the model, the purchase price lies between EUR 33.400 and EUR 41.000.

Model	Alfa Romeo Spider
Cubic capacity/cylinder	2198 cc/4 cylinder
hp/kW	185/136
Construction period	from 2006
Number of produced cars	–

Alfa Romeo Spider

In addition to a sound system featuring a CD player, the Alfa Romeo Spider is of course serially equipped with electrically controlled adjustable and heated exterior mirrors, a muffler featuring two pipes each on the left and on the right, air-conditioning, a leather steering wheel that is vertically and axially adjustable, as well as light alloy wheels. For safe driving, the car is equipped with front and side air bags as well as a knee air bag on the driver's side.

Model	Alfa Romeo Spider
Cubic capacity/cylinder	3195 cc/6 cylinder
hp/kW	260/191
Construction period	from 2006
Number of produced cars	–

Alfa Romeo GTV Edizione Sportiva

In 2001 Alfa Romeo introduced the special model GTV Edizione Sportiva. The model that was only offered in the noble colour "Black Metallic" featured high-quality leather interior equipment, a navigation system, as well as extremely wide tyres. In addition to the 150-hp 2.0 T.Spark version an even more vicious alternative was available, the 3.0 V6 24 VL.

Model	Alfa Romeo GTV
Cubic capacity/cylinder	2559 cc/6 cylinder
hp/kW	218/160
Construction period	2001
Number of produced cars	500

Alfa Romeo GTV

To make the elegant, wedge-shaped GTV Coupé even more vicious and faster, Alfa Romeo's technical engineers equipped this car as of 2003 with a 3.2-litre V6 engine. Up to that point, only drivers of the GTA class benefited from this 24-valve engine – now its 240 hp enabled the two-door GTV to achieve a top speed of 255 km/h.

Model	Alfa Romeo GTV
Cubic capacity/cylinder	3179 cc/6 cylinder
hp/kW	240/177
Construction period	2003 – 2005
Number of produced cars	–

Alfa Romeo GTV

In contrast to the early GTV models, as of 2003 the driver seat was placed lower and positioned more ergonomically. In addition, the cockpit received a newly designed centre console. Also new and completely in line with the prevailing trend was the possibility to order a radio-navigation system with an integrated CD player at a surcharge.

Model	Alfa Romeo GTV
Cubic capacity/cylinder	3179 cc/6 cylinder
hp/kW	240/177
Construction period	2003–2005
Number of produced cars	–

Alfa Romeo GT 1.8 TS

The year 2004 provided the fans of the Alfa Romeo brand with a new passion – the Alfa Romeo GT. With the introduction of this model, the Milan avant-garde brand is offering an extremely sporty four-seater available in the comfort lines Progression (comfort version with climate control, etc.) and Distinctive (deluxe version with leather interior).

Model	Alfa Romeo GT 1.8 TS
Cubic capacity/cylinder	1747 cc/4 cylinder
hp/kW	140/103
Construction period	from 2004
Number of produced cars	–

Alfa Romeo 8c Competizione

This two-seater prototype, designed by Centro Stile Alfa Romeo, should become a real icon of the brand – its bodywork construction consists of carbon fibre. At 190 centimetres, the vehicle is very wide and slightly reminiscent of the legendary 33 series Stradale. However, it is considerably faster with a maximum speed of at least 300 km/h.

Model	Alfa Romeo 8c Competizione
Cubic capacity/cylinder	4244 cc/8 cylinder
hp/kW	400/294
Construction period	2005
Number of produced cars	unique piece

Alfa Romeo 8c Competizione

Under the bonnet of the 8c Competizione sits a 4.2-litre V8 engine with two overhead camshafts for each row of cylinders. The engine, which is supported by a compressor, delivers more than 400 hp at 7000 revolutions and transfers this power via a manual six-step gear to the rear wheels. The acceleration from 0 to 100 km/h lasts, by the way, only 4.5 seconds.

Model	Alfa Romeo 8c Competizione
Cubic capacity/cylinder	4244 cc/8 cylinder
hp/kW	400/294
Construction period	2005
Number of produced cars	unique piece

Alfa Romeo Brera

Alfa Romeo composed the sporty elegance of the Brera in the typical form of this brand – in the design of the prototype the classical Alfa Romeo sign (also known as "Scudetto") was skilfully included in the layout of the front section. The gull-wing doors, which caused quite a stir in the concept car were, however not featured in the serial model.

Model	Alfa Romeo Brera
Cubic capacity/cylinder	4244 cc/8 cylinder
hp/kW	400/294
Construction period	2002
Number of produced cars	unique piece

Alfa Romeo Brera

Alfa Romeo continues with the Brera Coupé the tradition of its other famous models like the 1900 SS, the Giulietta Sprint or the Alfetta. The 441-centimetre long and 137-centimetre high coupé is offered in several output versions including an all-wheel model. Depending on its engine size, its maximum speed lies between 222 and 240 km/h.

Model	Alfa Romeo Brera
Cubic capacity/cylinder	2198 cc/4 cylinder
hp/kW	185/136
Construction period	from 2006
Number of produced cars	–

Alfa Romeo Brera

In the middle of the year 2006 followed the third engine variation of the Alfa Romeo Brera, a five-cylinder turbo-diesel engine with Multijet-injection. In addition to a basic model, the top version "Sky View" with a big panorama glass roof is also available. The sports car designed by Giorgetto Giugiaro – a so-called 2 + 2-seater – costs depending on its equipment between EUR 32,200 and EUR 39,800.

Model	Alfa Romeo Brera
Cubic capacity/cylinder	2387 cc/5 cylinder
hp/kW	200/147
Construction period	from 2006
Number of produced cars	–

Antas V8

The Italian company Faralli & Mazzanti, which has been specialised for many years in the restoration of noble old-timers and the upgrading of exclusive everyday cars, today also produces the Antas V8, a self-designed and manually constructed car. Under the modern interpretation of a classic design exterior, incidentally, there is a Maserati engine – a power source that allows the Antas V8 to reach a top speed of 270 km/h.

Model	Antas V8
Cubic capacity/cylinder	4719 cc/8 cylinder
hp/kW	310/228
Construction period	from 2004
Number of produced cars	–

Antas V8

According to information provided by the manufacturer, the Antas V8 is a high-caliber top-class touring car of which all construction steps are documented on a DVD for the proud owner. The DVD is handed over together with the vehicle documents and a heap of photographs in a blue velvet cover that features a silver badge with the engraved chassis number!

Model	Antas V8
Cubic capacity/cylinder	4719 cc/8 cylinder
hp/kW	310/228
Construction period	from 2004
Number of produced cars	–

Bertone Birusa

Bertone has already been co-operating with BMW since 1960, and many of the coupés made in Bavaria wore and continue to wear the handwriting of this Italian master coach maker. With the Birusa Bertone created a show car based on the BMW Z8, which was introduced to the general public at the Geneva Motor Show in 2003. Incidentally, the gull-wing doors of the Birusa that are made of carbon are electrically lifted!

Model	Bertone Birusa
Cubic capacity/cylinder	4941 cc/8 cylinder
hp/kW	400/294
Construction period	2003
Number of produced cars	unique piece

Bertone Birusa

At first sight, the Birusa, with its relatively clear and understated design lacks some classical car features like switches and other operating elements. These were deliberately omitted, because Bertone provided the Birusa with a voice activated system. In addition, the equipment of the concept car also includes an infrared night vision system and "flexible headlights" (that automatically adjust their brightness to their surroundings)

Model	Bertone Birusa
Cubic capacity/cylinder	4941 cc/8 cylinder
hp/kW	400/294
Construction period	2003
Number of produced cars	unique piece

Bizzarrini GTS

At the moment, the Italian company VGM Motors is working on a Bizzarrini project. It owns the trademark rights of the sports car manufacturer previously situated in Livorno and would like to launch a Bizzarrini car again on the market. A prototype of the GTS model already exists that is almost ready for serial production. According to factory specifications, it can reach a maximum speed of 360 km/h.

Model	Bizzarrini GTS
Cubic capacity/cylinder	4398 cc/8 cylinder
hp/kW	550/470
Construction period	from 2006
Number of produced cars	–

Cizeta Fenice TT Spyder

The Moroder company, that was reactivated in 2005 would like to continue the tradition of the V16T built in 1988 with the Cizeta Fenice TT Spyder. The Fenice – of which a yellow prototype can be regularly seen at exclusive car shows – is made exclusively by order. By the way, as opposed to the general practise, the 16-cylinder engine placed in the rear is inserted transversely rather than longitudinally.

Model	Cizeta Fenice TT Spyder
Cubic capacity/cylinder	5995 cc/16 cylinder
hp/kW	540/397
Construction period	from 2005
Number of produced cars	–

Covini C6W

One thing is for sure – everybody will certainly turn to stare at the C6W built by the Italian manufacturer Covini. This is because the low-riding super sports car moves on six wheels! This has only been tried once before in the sports car world when the British company Panther tried to establish as similar car, the Panther Six in the 1980s, unsuccessfully however.

Model	Covini C6W
Cubic capacity/cylinder	4200 cc/8 cylinder
hp/kW	380/279
Construction period	from 2006
Number of produced cars	–

Covini C6W

Covini already had the idea of creating a six-wheeled sports car in 1974. Following 30 years of meticulous work, it has finally become a reality. Now the serial production of the car with a top speed of 300 km/h and a long wheel base (275 centimetres) is expected to commence. The bodywork exists, incidentally, of carbon, and the engine technology is derived from Audi. The manufacturer hopes to be able to build and sell about six to eight cars per year.

Model	Covini C6W
Cubic capacity/cylinder	4200 cc/8 cylinder
hp/kW	380/279
Construction period	from 2006
Number of produced cars	–

Ferrari Enzo Ferrari

The model "Enzo Ferrari" (internally known as "FX") was only delivered to select customers who already called a Ferrari their own. The selling price of the sports car with a top speed of 355 km/h was around EUR 612.000. Originally, the Enzo had been planned as a limited series of 349 units. However, given the great demand, the manufacturer produced another 50 units.

Model	Ferrari Enzo Ferrari
Cubic capacity/cylinder	5998 cc/12 cylinder
hp/kW	660/485
Construction period	2002 – 2004
Number of produced cars	399

Ferrari FXX

The FXX model based on the "Enzo Ferrari" type is not street legal! Yet the lucky owners that are hand picked by the manufacturer can nevertheless step on its gas pedal, [illegible] price (approx. EUR [illegible] million) of the racer with a top speed of 350 km/h include driver's training on interna[illegible]

Model	Ferrari FXX
Cubic capacity/cylinder	6262 cc/12 cylinder
hp/kW	800/588
Construction period	2005 – 2006
Number of produced cars	30

Ferrari 612 Scaglietti

The body of the Scaglietti (successor of the 456 model) is completely made of aluminium and therefore combines high stability with low weight – compared to its predecessor it weighs about 60 kilograms less. To guarantee optimum weight distribution, the V12 engine with a strong torque is placed way at the back of the engine compartment.

Model	Ferrari 612 Scaglietti
Cubic capacity/cylinder	5748 cc/12 cylinder
hp/kW	540/397
Construction period	from 2003
Number of produced cars	–

Ferrari F 430 Coupé

With the F 430, Ferrari introduced a new vehicle generation to the market. The design and the technology of the new eight-cylinder engine were decisively influenced by the experience of the in-house Formula 1 department "Scuderia of Ferrari". This included technology solutions that were newly interpreted for the sporty street use and accordingly adapted.

Model	Ferrari F 430 Coupé
Cubic capacity/cylinder	4308 cc/8 cylinder
hp/kW	490/360
Construction period	from 2004
Number of produced cars	–

Ferrari F 430 Spider

Like the F 430 Coupé, the F 430 Spider also received a hypermodern engine with a flat crankshaft. The inspiration for this new engine, which achieves a maximum speed of more than 310 km/h, derives directly from Ferrari's racing models. The acceleration is equally impressive – the F 430 needs just 4.1 seconds to accelerate from 0 to 100 km/h.

Model	Ferrari F 430 Spider
Cubic capacity/cylinder	4308 cc/8 cylinder
hp/kW	490/360
Construction period	from 2005
Number of produced cars	–

Ferrari 599 GTB Fiorano

The 599 GTB Fiorano was named after the test track on which Ferrari develops its racing and street sports cars. It features a great number of technically very innovative solutions – for instance, the newest gear technology derived from the Formula 1. This is the only way to enable the switch process to be carried out at less than 100 milliseconds (!).

Model	Ferrari 599 GTB Fiorano
Cubic capacity/cylinder	5999 cc/12 cylinder
hp/kW	620/456
Construction period	from 2006
Number of produced cars	–

Lamborghini Murciélago

Faithful to its long-standing tradition, the Lamborghini has also named the Murciélago model after a fighting bull. The car with a top speed of 330 km/h is a two-seater coupé whose doors swing upwards to open. It features a mid-engine with a V12 design and with the engine operating band typical of Lamborghini – the gear transmission is located before the engine, behind it lies the differential as well as the permanent four-wheel drive.

Model	Lamborghini Murciélago
Cubic capacity/cylinder	6192 cc/12 cylinder
hp/kW	580/426
Construction period	from 2001
Number of produced cars	–

Lamborghini Murciélago Special Edition

On the occasion of the 40th anniversary of Lamborghini, the manufacturer presented in 2003 a special series of the Murciélago model. It was distinguished by pearlescent paint and differed from the serial version by some modifications of the interior equipment and the external appearance. It also featured a nice gesture by the manufacturer – every car included leather-covered suitcase made of carbon!

Model	Lamborghini Murciélago Special Edition
Cubic capacity/cylinder	6192 cc/12 cylinder
hp/kW	580/426
Construction period	2003
Number of produced cars	50

Lamborghini Murciélago Roadster

In 2003, the Murciélago Roadster was already shown as a concept car at the Motor Show in Detroit. In the meantime, the serial production of the dream car has started. In developing this vehicle, the designer Luc Donckerwolke, who has already designed the coupé, not only "cut off" the roof of the basic model, but he actually gave the super sports car an independent autonomous appearance.

Model	Lamborghini Murciélago Roadster
Cubic capacity/cylinder	6192 cc/12 cylinder
hp/kW	580/426
Construction period	from 2001
Number of produced cars	–

Lamborghini Murciélago LP 640

At the Geneva Motor Show of 2006 Lamborghini celebrated the world première of another dream car – the Murciélago LP 640. The term "LP" refers, incidentally, to the engine position ("longitudinale posteriore"). In designing the LP 640 (maximum speed 340 km/h), the designers of the Centro Stile Lamborghini adhered to the principles of purism, sportiness and functionality – resulting in an aggressive appearance.

Model	Lamborghini Murciélago LP 640
Cubic capacity/cylinder	6496 cc/12 cylinder
hp/kW	640/471
Construction period	from 2006
Number of produced cars	–

Lamborghini Gallardo

The Gallardo presented in 2003 is "the world's most beautiful car". This was the verdict of an international jury, consisting of designers, automobile historians and art historians as well as professional journalists. "This honour received by us", said the president of Lamborghini, "is not only very important for the Gallardo, but also for all employees of Automobili Lamborghini."

Model	Lamborghini Gallardo
Cubic capacity/cylinder	4961 cc/10 cylinder
hp/kW	520/382
Construction period	from 2003
Number of produced cars	–

Lamborghini Gallardo SE

One thing is probably certain: the vehicles that were put on the market in 2005 as a Gallardo Special edition are by now already rare collector's items. Compared to the standard version, the Gallardo SE features a transparent engine cover – technology becomes visible! The customers were given a choice of several two-colour exterior paints harmoniously matched to the likewise two-tone interior.

Model	Lamborghini Gallardo SE
Cubic capacity/cylinder	4961 cc/10 cylinder
hp/kW	520/382
Construction period	2005–2006
Number of produced cars	250

Lamborghini Gallardo Spyder

The Gallardo Spyder is almost ready for its market launch. Even before this date, the open model already has a track record of unusual success. For the complete annual production of 800 units there are already orders at the 86 Lamborghini traders around the globe! To add individual accents to the car, some interesting extras can be ordered – for example, a video camera in the rear spoiler that facilitates the parking process.

Model	Lamborghini Gallardo Spyder
Cubic capacity/cylinder	4961 cc/10 cylinder
hp/kW	520/382
Construction period	from 2006
Number of produced cars	–

Lancia Fulvia Coupé Concept

40 years after the debut of the sporty Fulvia, Lancia reinterpreted its layout once again – resulting in a modern coupé whose dimensions are almost equal to the original. In spite of the use of modern materials such as Alcantara, etc. the interior is also designed in the style of the 1960s and 1970s.

Model	Lancia Fulvia Coupé Concept
Cubic capacity/cylinder	1800 cc/4 cylinder
hp/kW	140/103
Construction period	2003
Number of produced cars	unique piece

Maserati Quattroporte

Maserati already built in the past various flagships under the model name Quattroporte, which were always distinguished by their aggressive lines as well as their prestigious dimensions. According to a company press release, the Quattroporte unites typically Italian values like a distinctive sense of beauty, cheerful enjoyment of life, reserved elegance, excellent mechanics and attractive exclusivity.

Model	Maserati Quattroporte
Cubic capacity/cylinder	4244 cc/8 cylinder
hp/kW	400/294
Construction period	from 2003
Number of produced cars	–

Maserati Gran Sport

In the 1950s, Maserati already introduced to the market a "Gran Sport Model", the Type A6 G. The new Gran Sport, another elegant coupé, continues the tradition of this series. It is created for all those in search of an especially spirited driving pleasure (maximum speed 290 km/h, acceleration from 0 to 100 km/h in just five seconds).

Model	Maserati Gran Sport
Cubic capacity/cylinder	4244 cc/8 cylinder
hp/kW	400/294
Construction period	from 2005
Number of produced cars	–

Maserati Gran Sport Spyder

The Gran Sport Spyder, which celebrated its world première in 2005 at the International Motor Show in Frankfurt, is the open alternative to the successful coupé variation. The car, which replaces the Spyder built up to that date, is based on the technology of the current coupé and is also available with a Cambiocorsa gear if requested.

Model	Maserati Gran Sport Spyder
Cubic capacity/cylinder	4244 cc/8 cylinder
hp/kW	400/294
Construction period	from 2006
Number of produced cars	–

Maserati Coupé

According to factory specifications, the Maserati Coupé consists of a thrilling sports car with four full seats that is very well suited for long trips. Its unmistakably Italian elegance is coupled with demanding technical solutions – the particularly compact and light V8 engine of this model has a structure that is normally only found in pure-bred racing engines.

Model	Maserati Coupé
Cubic capacity/cylinder	4244 cc/8 cylinder
hp/kW	390/287
Construction period	from 2005
Number of produced cars	–

Maserati MC 12

With the "Maserati Corse 12 cylinder" Maserati presents a vehicle of the FIA-GT class that, according to the regulations of the Fédération Internationale de l'Automobile (FIA) must be derived from a street model and of which at least 25 units have been produced. Maserati developed the design of the MC 12 together with Giugiaro based on the latest findings in the wind tunnel.

Model	Maserati MC 12
Cubic capacity/cylinder	5998 cc/12 cylinder
hp/kW	630/465
Construction period	2005–2006
Number of produced cars	25

Maserati Birdcage

With the Birdcage, Maserati introduced a car concept that is only 109 centimetres high and 202 centimetres wide and features 20-inch diameter front tyres and 22-inch diameter rear tyres. It supposedly achieves a maximum speed of 400 km/h. The bodywork of the race car consists of carbon fibre material and rests on a chassis derived from the Maserati MC 12. One searches in vain for the outside mirrors of the Birdcage – they have been replaced by cameras!

Model	Maserati Birdcage
Cubic capacity/cylinder	6000 cc/12 cylinder
hp/kW	700/515
Construction period	2005
Number of produced cars	unique piece

Pagani Zonda C 12

The sports car brand Pagani, headquartered in Italy, was founded in 1992 by the Argentine national Horacio Pagani. Pagani had previously worked for Lamborghini and Ferrari – his specialisation is carbon processing. The insight gained from his experience was now used for the creation of a super sports car that should bear his name as a model denomination – the prototype Pagani Zonda C 12 debuted in 1999.

Model	Pagani Zonda C 12
Cubic capacity/cylinder	5987 cc/12 cylinder
hp/kW	394/290
Construction period	1999–2002
Number of produced cars	5

Pagani Zonda C 12 S

It is known that Horacio Pagani was supported in the development of his sports car by the former Formula 1 driver Juan Manuel Fangio. Originally, the two-seater mid-engine coupé [illegible] the model name Pagani Fangio F1. However, following the death of Fangio in 1995 this idea was dropped. In contrast to the C 12, the C 12 S is equipped with a bigger engine.

Model	Pagani Zonda C 12 S
Cubic capacity/cylinder	7010 cc/12 cylinder
hp/kW	550/404
Construction period	1999–2003
Number of produced cars	15

Pagani Zonda C 12 S

When the first Pagani cars were offered for sale in 1999, the sports car world was enriched with a vehicle that was right away among the most exclusive. For the price of EUR 300,000, customers were assured that the annual production would not exceed 25 units. Another reassuring fact – the Zonda was powered by a Mercedes-Benz AMG engine allowing it to reach at least 290 km/h.

Model	Pagani Zonda C 12 S
Cubic capacity/cylinder	7010 cc/12 cylinder
hp/kW	550/404
Construction period	1999 – 2003
Number of produced cars	15

Pagani Zonda C 12 S 7.3

As the model name already indicates, a once more enlarged V12 engine is at work under the bonnet of this variation. To better utilise the torque of the AMG engine (at 4050 rpm 750 Newton metres are already available), the front part of the carbon fibre body was slightly reworked in favour of the aerodynamics.

Model	Pagani Zonda C 12 S 7.3
Cubic capacity/cylinder	7291 cc/12 cylinder
hp/kW	555/408
Construction period	2003 – 2004
Number of produced cars	–

Pagani Zonda C 12 S 7.3

With the launch of the C 12 S 7.3 model, the Zonda was available for the first time also as a roadster in addition to the usual coupé version. The dimensions of both versions are nearly identical and consisted of a length of 439.5 centimetres, a width of 205.5 centimetres, and a height of 115.1 centimetres. In terms of maximum speed, the car also operated at a high level – a breathtaking 340 km/h!

Model	Pagani Zonda C 12 S 7.3
Cubic capacity/cylinder	7291 cc/12 cylinder
hp/kW	555/408
Construction period	2003 – 2004
Number of produced cars	–

Pagani Zonda C 12 S 7.3

A few moves are all it takes to turn the Zonda roadster once again into a closed means of transportation. After driving the car with open roof, all one has to do is to insert the small roof middle section (resembling a Targa top). Nevertheless, whether open or closed – the currently fastest street-legal super sports car always cuts a fine figure.

Model	Pagani Zonda C 12 S 7.3
Cubic capacity/cylinder	7291 cc/12 cylinder
hp/kW	555/408
Construction period	2003 – 2004
Number of produced cars	–

Pagani Zonda F

The latest vehicle of the Pagani brand carries the model name Zonda F. Thus the Italian manufacturer remained true to its principle of naming the super car after the wind blowing in the South American Andes. Now, after all, the letter "F" commemorates (as opposed to the previous plan) the late racing driver Fangio.

Model Pagani Zonda F
Cubic capacity/cylinder 7291 cc/12 cylinder
hp/kW 602/443
Construction period from 2005
Number of produced cars –

Pagani Zonda F

In spite of its usual appearance, the Zonda F profits from much technical advancement that cannot be recognised at first sight. This includes ceramic brakes – they are the only ones able to securely delay the car that is by now equipped with 602 hp. Today, the acceleration to the 100-km/h mark amounts to 3.6 seconds, while the maximum speed is 345 km/h.

Model Pagani Zonda F
Cubic capacity/cylinder 7291 cc/12 cylinder
hp/kW 602/443
Construction period from 2005
Number of produced cars –

Spyker C8 Spyder

In 2000, the businessman Victor R. Muller revived the Dutch deluxe brand Spyker. At the Motor Show in Birmingham he presented with the C8 a deluxe sports car which was intended to continue the tradition of Spyker cars. The equation worked out – in late 2005, when the Spyker production started, 191 orders had already been placed.

Model Spyker C8 Spyder
Cubic capacity/cylinder 4172 cc/8 cylinder
hp/kW 400/294
Construction period from 2005
Number of produced cars –

Spyker C8 Spyder T

The Spyker C8 Spyder T celebrated its debut in 2005 at the Concours d'Elegance in Pebble Beach, California. The "T" in the model name refers to the upgraded turbo version – compared to the C8, the C8 T is equipped with an engine that was upgraded from the original 400 hp to 525 hp. Incidentally, it is based on the 4.2-litre V8 engine by Audi!

Model	Spyker C8 Spyder T
Cubic capacity/cylinder	4172 cc/8 cylinder
hp/kW	525/386
Construction period	from 2005
Number of produced cars	–

Spyker C8 Laviolette

Just like the Spyder version, a 4.2-litre Audi V8 engine powers the Laviolette coupé, presented for the first time in 2001.

Model	Spyker C8 Laviolette
Cubic capacity/cylinder	4172 cc/8 cylinder
hp/kW	450/335
Construction period	from 2005
Number of produced cars	–

Spyker C8 Laviolette

The Laviolette two-seater coupé is alternatively powered by a 405 hp or a 456 hp (twin turbo) Audi V8 engine. The attention-getting glass roof of the car with a top speed of 300 km/h is reminiscent of an aeroplane cockpit and is, incidentally, a homage to the brand's history – during the First World War, Spyker built pursuit planes.

Model	Spyker C8 Laviolette
Cubic capacity/cylinder	4172 cc/8 cylinder
hp/kW	405/298
Construction period	from 2005
Number of produced cars	–

Spyker C8 Double 12

A chassis extension forms the basis for the highly competitive C8 Double 12. Only by this feat it becomes possible to implant a six-litre W12 engine into the car. With this engine (630 hp), the Double 12 advanced into a sports car for long distance races. The longer rear deck version is a little more civilised – it is dominated by a 405 hp Audi V8 engine.

Model	Spyker C8 Double 12
Cubic capacity/cylinder	4172 cc/8 cylinder
hp/kW	405/298
Construction period	from 2005
Number of produced cars	–

Spyker C 12 La Turbie

Thanks to a modern chassis construction (aluminium space frame), the Spyker tops the scale at an average of only 1295 kilograms. In connection with the aerodynamic bodywork, this results in excellent acceleration values. Depending on the engine output, the 100-km/h mark is reached arrived after only four seconds – while the maximum speed is approximately 320 km/h.

Model	Spyker C12 La Turbie
Cubic capacity/cylinder	5998 cc/8 cylinder
hp/kW	500/373
Construction period	from 2006
Number of produced cars	–

Koenigsegg CC 8 S

Sweden of all places, a country with winters abundant in snow, is the place of origin of the Koenigsegg, a super sports car for enthusiasts who would like to own something different than a Ferrari or Lamborghini. Christian von Koenigsegg established his sports car manufacturing company in 1993 in Angelholm in southern Sweden and four years later already presented the first prototypes of the "CC" model built in series as of 2002.

Model	Koenigsegg CC 8 S
Cubic capacity/cylinder	4600 cc/V8 cylinder
hp/kW	655/482
Construction period	from 2002
Number of produced cars	–

Koenigsegg CC 8 S

A shape to fall in love with! The stars and starlets probably also felt this way when Koenigsegg in 1997 presented them with a prototype model of the car on the occasion of the film festival in Cannes. Who wanted to own this object of desire, had to have plenty of patience – since the first car, which in the meantime was called CC 8 S, would only be handed over to its proud owner in 2002 at the Geneva International Motor Show.

Model	Koenigsegg CC 8 S
Cubic capacity/cylinder	4600 cc/V8 cylinder
hp/kW	655/482
Construction period	from 2002
Number of produced cars	–

Koenigsegg CC 8 S

In 2001, the Scandinavian automotive specialist publications were already certain that the title "Car of the year" was beyond any doubt due to the Koenigsegg CC 8 S. The sports car could also secure the honour of being the "best Swedish design". In contrast to other, often very modestly equipped, sports cars, the CC 8 S with a top speed of 390 km/h, is abundant in luxuries – anything is possible, even an especially designed luggage set!

Model	Koenigsegg CC 8 S
Cubic capacity/cylinder	4600 cc/V8 cylinder
hp/kW	655/482
Construction period	from 2002
Number of produced cars	–

Koenigsegg CCR

With the Koenigsegg CCR, which the Swedish sports car manufacture showed in 2004 at the Geneva Motor Show, debuted a street sports car whose speedometer needle only came to a halt right before the 400 km/h.

The secret of the CCR consists of a considerable power boost and numerous technical modifications to the Ford engine (including a compressor that exerts a pressure of 1.2 bar on the cylinders).

Model	Koenigsegg CCR
Cubic capacity/cylinder	4700 cc/V8 cylinder
hp/kW	816/600
Construction period	from 2004
Number of produced cars	–

Koenigsegg CCR

To enable comfortable access to the CCR (whose upholstery is made of the NASA material Tempur), an unusual swivel mechanism was contrived which allows the doors to rotate at first sideways and then up. Who would like to make do without the contortions, should better drive the car with an open top – the low roof can be dismantled with a few moves and stowed away under the front bonnet.

Model	Koenigsegg CCR
Cubic capacity/cylinder	4700 cc/V8 cylinder
hp/kW	816/600
Construction period	from 2004
Number of produced cars	–

Koenigsegg CCR

As opposed to Ferrari, the shape of the CCR did not originate at a design studio, but in-house at Koenigssegg itself.The monocoque, which consists of 21 layers of carbon, rather resembles the cockpit of a fighter jet and is the result of extensive tests in the wind tunnel. This effort has paid off – in the first test trips already the CCR achieved a maximum speed of 389 km/h.

Model	Koenigsegg CCR
Cubic capacity/cylinder	4700 cc/V8 cylinder
hp/kW	816/600
Construction period	from 2004
Number of produced cars	–

Koenigsegg CCX

With the model CCX (Competition Coupé X) Koenigsegg would like to commemorate 1996 – the year in which the first ready to drive prototype of the car left the factory halls. Compared to the CC and CCR models, the CCX design looks rather subdued and less aggressive – to correspond with American licensing regulations, the bumper areas had to be changed in the front and rear.

Model	Koenigsegg CCX
Cubic capacity/cylinder	4700 cc/V8 cylinder
hp/kW	806/593
Construction period	from 2006
Number of produced cars	–

Koenigsegg CCX

The delivery time for a sports car by Koenigsegg amounts to about eight to twelve weeks. Depending on its equipment and requested extras the driving pleasure begins at EUR 458,000 (plus taxes). At the completion of the contract, the Swedish noble brand already requests a deposit of 25 percent and points out in a friendly fashion that the balance should be paid in full two weeks prior to delivery.

Model	Koenigsegg CCX
Cubic capacity/cylinder	4700 cc/V8 cylinder
hp/kW	806/593
Construction period	from 2006
Number of produced cars	–

Quercianella

The Swedish company Laboratorio Bebi wants to join the ranks of car manufacturers and will soon launch a sports car with a top speed of 300 km/h, the Quercianella. The mid-engine vehicle should become available in several output levels (380 hp to 525 hp). It is made of lightweight construction materials and weighs only a modest 525 kilograms.

Model	Quercianella
Cubic capacity/cylinder	2400 cc/8 cylinder
hp/kW	380/279
Construction period	from 2007
Number of produced cars	–

Saab 9-3 "20 years edition"

With the introduction of the current 9-3 generation in the summer of 2003, Saab switched to the Magna Steyr experts in the vicinity of Graz (Austria) for the assembly of the vehicle – until 2003 they had been commissioning the Finnish supplier Valmet with the manufacture of the convertible. Customers can choose among the by now extensive range of models.

Model	Saab 9-3 "20 years edition"
Cubic capacity/cylinder	2800 cc/6 cylinder
hp/kW	250/184
Construction period	from 2003
Number of produced cars	–

Saab Aero X

The world première of a powerful concept car determined the display of the Swedish automobile manufacturer Saab at the Geneva Motor Show of 2006. The design of the "Aero X" was inspired by aeronautics, resulting in a very special two-seater sports coupé. Immediately noticeable is the absence of doors as well as the A column, since the Aero X features a cockpit similar to that of a jet plane.

Model	Saab Aero X
Cubic capacity/cylinder	2800 cc/6 cylinder
hp/kW	400/294
Construction period	2006
Number of produced cars	individual piece

Saab Aero X

The Saab concept car Aero X (4,675 milimetre in length) moves in the front on tyres sized 265/30, and in the rear on tyres measuring 325/25. It accelerates from 0 to 100 km/h in only 4.9 seconds and reaches an electronically controlled maximum speed of 250 km/h. The power transmission takes place, incidentally, via a seven-gear automatic transmissions with a dual clutch.

Model	Saab Aero X
Cubic capacity/cylinder	2800 cc/6 cylinder
hp/kW	400/294
Construction period	2006
Number of produced cars	unique piece

RInspeed Advantige R One

In developing the R One, Rinspeed aimed to create a "down-to-earth" sports car that not only provides plenty of driving pleasure, but is also distinguished by a unique design and above all by its environmental friendliness. Therefore, the R One with a top speed of 205 km/h can be fuelled with "Kompogas", a gas generated from organic waste.

Model	Rinspeed Advantige R One
Cubic capacity/cylinder	1795 cc/4 cylinder
hp/kW	120/88
Construction period	2001
Number of produced cars	–

Rinspeed Bedouin

In 2003, the Swiss "creative manufacturer" came out with a surprising and quite special attraction, the Rinspeed Bedouin. As an absolute novelty, this car can be transformed in less than ten seconds by pushing a button from a two-seater pickup truck into a sporty four-seater estate car with a four-wheel design.

Model	Rinspeed Bedouin
Cubic capacity/cylinder	3600 cc/6 cylinder
hp/kW	420/309
Construction period	2003
Number of produced cars	individual piece

Rinspeed Bedouin

With the Bedouin, Rinspeed wants to demonstrate that there remains much creative potential inherent in the category of multipurpose vehicles. Besides being a real master of metamorphosis, the Bedouin is the fastest natural gas vehicle in the world. However, its environmental friendliness does not negatively impact the driving pleasure. The car reaches a speed of 100 km/h after 5.9 seconds, while its regulated maximum speed is 250 km/h.

Model	Rinspeed Bedouin
Cubic capacity/cylinder	3600 cc/6 cylinder
hp/kW	420/309
Construction period	2003
Number of produced cars	individual piece

Rinspeed Senso

Together with several partners from the automotive industry, Rinspeed has once again created a very unusual vehicle – the concept car Senso. With its sensor technology it "senses" the driver, measures his/her biometric data and then positively influences the driver with patterns, colours, music and scents!

Model	Rinspeed Senso
Cubic capacity/cylinder	3200 cc/6 cylinder
hp/kW	250/184
Construction period	2005
Number of produced cars	unique piece

Rinspeed Senso

The Senso is based on advanced technology. This includes a camera that records the driving behaviour of the driver while a onboard calculator determines his/her its state of mind based on this data. On the basis of these values, the pilot then receives the "sensory-pleasing" impulses that transform him/her into a state of relaxed attention.

Model	Rinspeed Senso
Cubic capacity/cylinder	3200 cc/6 cylinder
hp/kW	250/184
Construction period	2005
Number of produced cars	unique piece

Rinspeed zaZen

Rinspeed in conjunction with Bayer Material Science developed the concept car "zaZen": a "body of light" on four wheels. It features a technical revolution in automobile construction, because the transparent rear window becomes a holographic luminous surface. Seemingly from nowhere, the third stoplight shines out of the apparently floating and lucid "hard top"!

Model	Rinspeed zaZen
Cubic capacity/cylinder	3824 cc/6 cylinder
hp/kW	355/261
Construction period	2006
Number of produced cars	unique piece

Rinspeed zaZen

The "zaZen" with a total weight of 1495 kilograms is powered by a six-cylinder Boxer engine by Porsche. To reach a speed of 100 km/h it requires 4.8 seconds, while its maximum speed amounts to 293 km/h. Should the "zaZen" be built as a small series, Rinspeed is considering offering it also with an environmentally friendly natural gas engine!

Model	Rinspeed zaZen
Cubic capacity/cylinder	3824 cc/6 cylinder
hp/kW	355/261
Construction period	2006
Number of produced cars	unique piece

Identity i1

The Spanish national José Ramón Fernández Franqueira began his engineering career first of all at Fiat Auto Spain, before establishing his own automobile manufacturing company in 2002. Franqueiras first creation, a high-calibre sports car, received a great amount attention the Geneva Motor Show in 2006. The model that was designed in co-operation with the design studio H2R, is expected to be launched soon under the name Identity i1.

Model	Identity i1
Cubic capacity/cylinder	–/8 cylinder
hp/kW	487/358
Construction period	from 2006
Number of produced cars	–

Identity i1

The only 1261 millimetre high Identity i1 will not be seen very often. According to the factory specifications, a maximum of 300 wide-wheel units will be produced and then the production ceased. The chassis of the fast two-seater is made, by the way, of composite materials. The acceleration from 0 to 100 km/h should at any rate be less than four seconds, Identity still has not released any information concerning its maximum speed.

Model	Identity i1
Cubic capacity/cylinder	–/8 cylinder
hp/kW	487/358
Construction period	from 2006
Number of produced cars	–

Tramontana

This is not a new concept. In the 1950s, there were already cars with two seats arranged behind each other – they were called micro cars. However, the Tramontana made in the Spanish Costa Brava is anything but a small car. The manufacturer, who equips the vehicle with a twelve-cylinder engine, refers to it as a combination of a jet aeroplane and Formula 1 racer.

Model	Tramontana
Cubic capacity/cylinder	5500 cc/12 cylinder
hp/kW	720/529
Construction period	from 2006
Number of produced cars	–

Tramontana

The Tramontana – a plaything for the rich and famous – became available for the first time in summer 2006. To maintain its exclusivity, annually a maximum of twelve of these crazy vehicles will be produced. The entrance level price is EUR 600,000. Since the manufacturer is willing to comply with special customer wishes, the upper price limit remains open.

Model	Tramontana
Cubic capacity/cylinder	5500 cc/12 cylinder
hp/kW	720/529
Construction period	from 2006
Number of produced cars	–

Cadillac Cien

With the aerodynamic Cien – Spanish for "100" – Cadillac celebrated in 2002 the 100th birthday of the brand. The sleek appearance of the Cien, inspired by the latest F-22-Stealth fighter planes, has created a new look for Cadillac. The Cien is equipped with a semiautomatic gear transmission, which is operated via electronic paddles mounted on the steering column.

Model	Cadillac Cien
Cubic capacity/cylinder	7500 cc/12 cylinder
hp/kW	750/551
Construction period	2002
Number of produced cars	unique piece

Cadillac Sixteen

Under the bonnet of the Sixteen there is an unequalled engine – with a bore of 105 millimetres and a piston stroke of 98 millimetres, the 16 cylinders dispose of the impressive cubic capacity of 13.6 litres. The engine provides a full 1000 hp with 6000 rpm. Due to its weight (approx. 1700 kilograms), the Sixteen reaches a maximum speed of 227 km/h.

Model	Cadillac Sixteen
Cubic capacity/cylinder	13,600 cc/16 cylinder
hp/kW	1000/735
Construction period	2002
Number of produced cars	unique piece

Cadillac Sixteen

In 2002 – on the occasion of its 100th anniversary – Cadillac exhibited a number of unusual show cars, including this luxurious 16 cylinder vehicle. With a wheel base of 355.6 centimetres and a length of 567.3 centimetres the Cadillac Sixteen cuts an impressive figure. Its silhouette is dominated by the long bonnet as well as the 24-inch wheels with tyre measuring P 265/40 R25.

Model	Cadillac Sixteen
Cubic capacity/cylinder	13,600 cc/16 cylinder
hp/kW	1000/735
Construction period	2002
Number of produced cars	unique piece

Cadillac XLR V

The XLR V is the latest offspring of the growing family of high-capacity vehicles by Cadillac. It combines luxury with strength, performance, personal comfort and style. The XLR V is equipped with the "Northstar V-8-SC engine" with compressor charging. Featuring a six-speed-automatic gear, it accelerates in less than 5 seconds from 0 to 100 km/h and has a high speed of more than 250 km/h.

Model	Cadillac XLR V
Cubic capacity/cylinder	4371 cc/8 cylinder
hp/kW	450/330
Construction period	from 2005
Number of produced cars	–

Cadillac XLR V

The XLR-V model, derived from the "Evoq" concept car presented in 1999, has a compact length of 451 centimetres. The XLR V belongs to the class of luxurious sport roadsters and is equipped with a stable, foldable metal hard top – an electrohydraulic system transforms the car at the touch of a button in only 20 seconds into an attractive roadster.

Model	Cadillac XLR V
Cubic capacity/cylinder	4371 cc/8 cylinder
hp/kW	450/330
Construction period	from 2005
Number of produced cars	–

Chevrolet Camaro Concept

With a combination of dramatic design and exciting performance, Chevrolet in 2006 took up once more the concept of the Camaro, one of the most popular sports coupés of all times, and redefined it for another fan generation. It aimed primarily to develop a brand-new version of the Camaro based on customer reactions to the design and technological innovations.

Model	Chevrolet Camaro Conc.
Cubic capacity/cylinder	5999 cc/8 cylinder
hp/kW	400/298
Construction period	2006
Number of produced cars	unique piece

Chevrolet Corvette C 6 Coupé

The Corvette of the sixth generation is available as a coupé or a convertible and constitutes the Europe's most popular Corvette model of all times. As a novelty, it became available as of 2006 for the first time with a six-speed-automatic gear with an electronically steered "paddle shift" that allows impressive acceleration values coupled with low fuel consumption.

Model	Chevrolet Corvette C 6 Coupé
Cubic capacity/cylinder	5967 cc/8 cylinder
hp/kW	405/298
Construction period	from 2004
Number of produced cars	–

Chevrolet Corvette C 6 Cabriolet

With a c_d (coefficient of drag) value of 0.28, the current Corvette is the most aerodynamic model of all times. Due to modern technological possibilities the shape and surfaces could be kept very smooth. For example, the keyless access system replaces the conventional mechanical handles on the doors and the hatch with solenoids and electronic actuators.

Model	Chevrolet Corvette C 6 Cabriolet
Cubic capacity/cylinder	5967 cc/8 cylinder
hp/kW	405/298
Construction period	from 2004
Number of produced cars	–

Chevrolet Corvette Z 06 Coupé

The Corvette coupé disposes of a removable roof, which can be stowed away in the boot. While the metal sheet of the roof is around 15 percent bigger than that of the previous models, it nevertheless weighs only 0.45 kilograms more. It is serially available in the colour of the car or optionally with a transparently tinted roof part.

Model	Chevrolet Corvette Z 06 Coupé
Cubic capacity/cylinder	7011 cc/8 cylinder
hp/kW	512/377
Construction period	from 2005
Number of produced cars	–

Chevrolet Corvette Z 06 Cabriolet

Because the Corvette has been conceived from the outset as an open car, in comparison to the coupé no compromises in terms of ride comfort, handling or the performance had to be made for the convertible. The Corvette convertible of the sixth generation has an electronic soft top, which can be opened or closed in only 10 seconds.

Model	Chevrolet Corvette Z 06 Cabriolet
Cubic capacity/cylinder	7011 cc/8 cylinder
hp/kW	512/377
Construction period	from 2005
Number of produced cars	–

Chevrolet SSR

This type of car can only exist in the USA – with the SSR, Chevrolet has succeeded in creating a more manageable and sporty version of a pick up, the car style that is very popular and usually quite bulky in the USA. The first design studies of the SSR (Super Sport Roadster) date back to the year 1999 – while in the winter of 2002/2003 the first serial production version was presented.

Model	Chevrolet SSR
Cubic capacity/cylinder	5967 cc/8 cylinder
hp/kW	405/298
Construction period	from 2003
Number of produced cars	–

Chevrolet SSR

The special feature of the Super Sport Roadster SSR is its ingenious power-retractable metal hardtop. When the roof is closed, the SSR driver sits enthroned in a small pick-up cabin – when open, this feature transforms the 486 centimetre long pickup truck in a two-seater roadster with plenty of storage space behind the seats. The maximum speed of the SSR amounts to 210 km/h.

Model	Chevrolet SSR
Cubic capacity/cylinder	5967 cc/8 cylinder
hp/kW	405/298
Construction period	from 2003
Number of produced cars	–

Chrysler Crossfire

After the merger of Daimler-Benz and Chrysler, the first joint product was the Chrysler Crossfire. In 2001 the two-seater coupé was introduced to the general public as a concept car almost ready for serial production – a year later the joint product was offered for retail.

Model	Chrysler Crossfire
Cubic capacity/cylinder	3199 cc/6 cylinder
hp/kW	218/160
Construction period	2000
Number of produced cars	unique piece

Chrysler Crossfire

In terms of technology, the Crossfire is nothing other than the continuation of the first series of the Mercedes-Benz SLK Roadster. This is where the technology was derived from (chassis and engine), while Chrysler provided the design of the body. However, the successful model is not built in the United States, but by the coach maker specialist Karmann in Osnabrück, Germany!

Model	Chrysler Crossfire
Cubic capacity/cylinder	3199 cc/6 cylinder
hp/kW	218/160
Construction period	from 2002
Number of produced cars	–

Chrysler Crossfire Roadster

At the Geneva Motor Show in 2004, Chrysler Group presented the open version of the Crossfire coupé. Under the bonnet of the car offering open driving pleasure also sits a 3.2-litre V6 engine. The top can be opened electrically and completely disappears after 22 seconds under a hard tonneau panel.

Model	Chrysler Crossfire Roadster
Cubic capacity/cylinder	3199 cc/6 cylinder
hp/kW	218/160
Construction period	from 2004
Number of produced cars	–

Chrysler Crossfire Roadster SRT 6

The Crossfire with the additional abbreviation SRT is the result of a technology transfer within the DaimlerChrysler Group. The SRT Organisation (Street and Racing Technology) – i.e. the racing department of the Group – provides the Crossfire as well as other models with even more sportiness. In plain English, the Crossfire engine was upgraded from 218 hp to 335 hp.

Model	Chrysler Crossfire Roadster SRT 6
Cubic capacity/cylinder	3199 cc/6 cylinder
hp/kW	335/246
Construction period	from 2004
Number of produced cars	–

Chrysler Crossfire SRT 6

The beefy strength of the SRT 6 is transferred via a five-step automatic transmission with optional manual operation ("Autostick") to the rear wheels and allows a maximum speed of 255 km/h. The SRT 6 accelerates from a standing position to the 100-km/h mark within 5.3 seconds – while the "standard" Crossfire without compressor support requires 6.5 seconds.

Model	Chrysler Crossfire SRT 6
Cubic capacity/cylinder	3199 cc/6 cylinder
hp/kW	335/246
Construction period	from 2004
Number of produced cars	–

Chrysler ME Four-Twelve

The ME Four-Twelve is absolutely the most impressive car ever introduced by Chrysler. Due to a brawny V12 engine, the acceleration from 0 to 100 km/h takes only three seconds, and whoever steps all the way on the gas pedal will see that the speedometer needle only comes to a halt at the 400 km/h mark.

Model	Chrysler ME Four-Twelve
Cubic capacity/cylinder	5998 cc/12 cylinder
hp/kW	850/625
Construction period	2003
Number of produced cars	unique piece

Chrysler Firepower

Based on the Dodge Viper, Chrysler created the impressive Firepower concept car, which the automotive press already hails as the successor of the Crossfire model. Thanks to its generous tyres (in the rear 335/30 R 20) the concept car already looks vigorous while standing still. It is indeed vigorous – requiring from 0 to 100 km/h in only 4.5 seconds, and with a maximum speed of 282 km/h.

Model	Chrysler Firepower
Cubic capacity/cylinder	6059 cc/8 cylinder
hp/kW	422/310
Construction period	2005
Number of produced cars	unique piece

Dodge Razor

The Dodge Razor concept car is the joint effort of the Chrysler Group Design Studios in Auburn Hill, Michigan and Carlsbad (California). The designers were inspired in their work by the classical European sports cars of the 1960s and favoured primarily a simple and realistic body design.

Model	Dodge Razor
Cubic capacity/cylinder	2400 cc/4 cylinder
hp/kW	250/187
Construction period	2001
Number of produced cars	unique piece

Dodge Razor

To keep the price of the Razor affordable – if it should [illegible] go into production – a chassis as well as chassis components from the DaimlerChrysler contingent were primarily used in this design effort. The engine that is equipped with a turbocharger enables a maximum speed of 225 km/h – theoretically, the driving pleasure could be realised at USD 14,500.

Model	Dodge Razor
Cubic capacity/cylinder	2400 cc/4 cylinder
hp/kW	250/187
Construction period	2001
Number of produced cars	unique piece

Dodge Viper SRT 10 Cabriolet

When in 2002 the first Viper Cabriolets of the third generation were released, at that point manufactured under the management of the DaimlerChrysler Group, the fascination with this ultimate American sports car remained unbroken. In the meantime, in addition to the hues of red and black the colour palette was enriched with yellow and other nuances as well.

Model	Dodge Viper SRT 10 Cabriolet
Cubic capacity/cylinder	8277 cc/10 cylinder
hp/kW	517/380
Construction period	from 2002
Number of produced cars	–

Dodge Viper SRT 10 Coupé

Finally, to the great pleasure of Viper fans, the long awaited SRT 10 Coupé was shown at the Geneva Motor Show of 2005. While all previous models were already known for absolute top performance, it was nevertheless worthwhile to re-examine the technical data – the acceleration to the 100-km/h mark is completed in four seconds, while the maximum speed amounts to 306 km/h!

Model	Dodge Viper SRT 10 Coupé
Cubic capacity/cylinder	8277 cc/10 cylinder
hp/kW	517/380
Construction period	from 2005
Number of produced cars	–

Dodge Charger RT

Similar to its his predecessors of the 1960s, the Charger is a car consciously designed for sportiness. The four-door sedan is offered in various engine variations and has 253 hp. Even more fun is provided by the SRT versions – i.e. those models that were provided with substantially more bite by the Chrysler racing department.

Model	Dodge Charger RT
Cubic capacity/cylinder	5654 cc/8 cylinder
hp/kW	345/254
Construction period	from 2005
Number of produced cars	–

Dodge Charger SRT 8 Super Bee

A maximum speed of 250 km/h is definitely a remarkable result for a heavy four-door sports sedan. This value can be reached by the Super Bee version of the Dodge Charger. The "Superbee" of today thus continues the tradition of the 1970s when Dodge released its first swarm of "Super Bees".

Model	Dodge Charger SRT8 Super Bee
Cubic capacity/cylinder	4430 cc/8 cylinder
hp/kW	345/318
Construction period	from 2006
Number of produced cars	–

Dodge Challenger Concept

The plan view "hip" and the black stripe on the bonnet are only some of the details used by Dodge to commemorate the Challenger Concept of the late 1960s. Professional journalists already speculate that the new edition with the retro design could become a reality soon as a standard model at a cost of USD 47.000.

Model	Dodge Challenger Concept
Cubic capacity/cylinder	6100 cc/8 cylinder
hp/kW	425/313
Construction period	2006
Number of produced cars	unique piece

Fisker Latigo

Henrik Fisker and Bernhard Koehler both have a great amount of experience in the automobile construction sector. The Danish national Fisker worked for a long time for Aston Martin and BMW, while his German colleague had worked for Ford. Now the two of them are producing high-calibre cars themselves in the USA. The technical base of the Fisker Latigo is the BMW 6 series coupé whose prototype can still be made out.

Model	Fisker Latigo
Cubic capacity/cylinder	6100 cc/10 cylinder
hp/kW	507/373
Construction period	from 2006
Number of produced cars	–

Fisker Tramonto

Anyone who puts a Mercedes-Benz SL 55 AMG between the hands of Fisker, receives for EUR 110.000 a new carriage which has almost nothing in common anymore with its original source base. This is because the Californian sports car manufacturer provides the SL with the silhouette of a fighter jet in the front and upgrades the interior as well as the engine. The result is a car with a top speed of 323 km/h that is unparalleled.

Model	Fisker Tramonto (Performance)
Cubic capacity/cylinder	5439 cc/8 cylinder
hp/kW	610/400
Construction period	from 2005
Number of produced cars	–

Fisker Tramonto

Anyone who would like to own a Fisker, must previously acquire the base car – in this case a Mercedes-Benz SL 55 AMG – and transport it to the USA. Here the roadster is thoroughly reworked in terms of appearance and technology. [illegible] not much left of the original. The result is a sports car with a top speed of 250 km/h ([illegible] version) which now carries the name Fisker Tramonto and accelerates in 4.5 seconds to 100 km/h.

Model	Fisker Tramonto (Standard)
Cubic capacity/cylinder	5439 cc/8 cylinder
hp/kW	493/323
Construction period	from 2005
Number of produced cars	–

Ford Thunderbird

Ford had started officially taking orders for the Thunderbird that was built again from 2002 on the 8th of January, 2001 – the day when the definitive serial model was introduced. The uniquely styled two-seater convertible whose rear wheels are driven by a V8 engine provides – like the Thunderbird of the 1950s – a feeling of automotive individuality and freedom.

Model	Ford Thunderbird
Cubic capacity/cylinder	3900 cc/8 cylinder
hp/kW	243/179
Construction period	2002–2004
Number of produced cars	–

Ford Thunderbird

As an extra equipment, a removable hard top is available for the Thunderbird which features – just like the previous models – classical "porthole windows". The interior of the two-seater also combines its romantic history with very expressive design and the comfort demanded by today's customers from a sporty vehicle.

Model	Ford Thunderbird
Cubic capacity/cylinder	3900 cc/8 cylinder
hp/kW	243/179
Construction period	2002 – 2004
Number of produced cars	–

Ford Mustang GT

While some automobile manufacturers are reviving the fifties and sixties in the shape of successful design studies, Ford is already offering a real retro classic car as a standard model in its portfolio – the Mustang. According to its classical construction design, this model has no independent rear suspension, but a rigid axle instead!

Model	Ford Mustang GT
Cubic capacity/cylinder	4606 cc/8 cylinder
hp/kW	304/224
Construction period	from 2003
Number of produced cars	–

Ford Mustang Cabrio 4.0

Starting in spring 2005, a Mustang convertible whose design is strongly reminiscent of the classical Mustang of the 1960s is sold again in the USA. Its standard engine is a 4.0-litre V6 engine with an overhead camshaft that provides a torque of 325 Newton metres. The top speed of the 477 centimetre long carriage amounts to 190 km/h.

Model	Ford Mustang Cabrio 4.0
Cubic capacity/cylinder	4009 cc/6 cylinder
hp/kW	213/157
Construction period	from 2005
Number of produced cars	–

Ford Mustang Shelby GT 500

"Supercharged" – i.e. supported by a compressor, the Mustang modified by Carroll Shelby reaches a substantial speed of 270 km/h. The considerable additional horse power provided to this model, featuring a rigid rear axis, allows it to accelerate from 0 to 100 km/h in only 4.7 seconds.

Model	Ford Mustang Shelby GT 500
Cubic capacity/cylinder	5409 cc/8 cylinder
hp/kW	482/355
Construction period	from 2006
Number of produced cars	–

Ford GT 40

The GT 40 – a modern contemporary sports car – is inspired by its predecessor model of the same name, which in the middle of the 1960s wrote a new chapter in the history of high performance vehicles in long distance races. Although the "new" version is a little longer and higher than the original, it incorporates all of its typical performance data.

Model	Ford GT 40
Cubic capacity/cylinder	5409 cc/8 cylinder
hp/kW	550/410
Construction period	from 2003
Number of produced cars	

Ford GT 40

The Ford GT 40 concept car presented in January 2002 at the Detroit Motor Show has been produced since 2003 as a serial model right on time for the 100th anniversary of the founding of the enterprise. The GT 40 enriches the sports car scene and complements the Ford's "Living Legends" series, which also includes the Thunderbird and Mustang models.

Model	Ford GT 40
Cubic capacity/cylinder	5409 cc/8 cylinder
hp/kW	550/410
Construction period	from 2003
Number of produced cars	–

Ford Shelby GR-1 Concept Car

The Ford Shelby GR-1 concept car was one of the vehicles that were especially admired at the Detroit engine show in 2005. It features a strong V10 all-aluminium engine and is equipped with many components that are used in the Ford GT. The polished aluminium body looks very realistic – a serial production of the GR-1 would be desirable.

Model	Ford Shelby GR-1 Concept Car
Cubic capacity/cylinder	6400 cc/10 cylinder
hp/kW	605/445
Construction period	2005
Number of produced cars	unique piece

Mosler MT 900

The just 400 centimetre-long and only 113 centimetre-high Mosler sports car is a native of sunny Florida and extremely rarely found on European streets. Nevertheless, it would be no problem to find maintenance and service for this exotic car on this continent – its gear is also used in the Porsche 911, while its engine technology should be familiar to every Chevrolet Corvette mechanic.

Model	Mosler MT 900
Cubic capacity/cylinder	5665 cc/8 cylinder
hp/kW	435/320
Construction period	from 2005
Number of produced cars	–

Panoz Esperante

Donald Panoz has first made a fortune with nicotine patches and subsequently invested in the construction of automobiles. Since 1999, the son of an Italian immigrant to the USA has been producing high-calibre top sports cars. Some of his unconventional models are used in racing (Le Mans), while others – like the Esperante – enrich the street scene as street sports cars.

Model	Panoz Esperante
Cubic capacity/cylinder	4601 cc/8 cylinder
hp/kW	309/227
Construction period	from 1999
Number of produced cars	–

Panoz Esperante GTLM Coupé

There are actually no great differences between the Esperante models, apart from their engine output: the sports cars are equipped with a V8 engine by Ford and transfer their power via a manual five-speed gearbox to the rear axle. The Esperante GTLM is supported, in addition, by a turbocharger and reaches a maximum speed of 289 km/h (Esperante = 249 km/h).

Model	Panoz Esperante GTLM Coupé
Cubic capacity/cylinder	4601 cc/8 cylinder
hp/kW	420/309
Construction period	from 2001
Number of produced cars	–

Panoz Esperante GTLM Cabriolet

Thanks to a light frame made of aluminium sections, the Esperante tops the scales at only 1115 kilograms and reaches the 100-km/h mark after five seconds. To maintain the exclusivity of a Panoz car, a maximum of 30 units are produced annually according to the manufacturer. The driving pleasure is available starting at USD 121,000.

Model	Panoz Esperante GTLM Cabriolet
Cubic capacity/cylinder	4601 cc/8 cylinder
hp/kW	420/309
Construction period	from 2002
Number of produced cars	–

Saleen S7

Saleen Inc. was established in 1983 by the Californian native Steve Saleen and is famous far beyond the USA for the most aggressive vehicle tuning (primarily Ford Mustang models). In addition, the company also produces pure-bred racers like the type S7R. Of this model, which received plenty of attention in 2001 at Le Mans, there is nowadays also a "tame" street version available for the considerable price of approximately USD 400,000.

Model	Saleen S7
Cubic capacity/cylinder	7000 cc/V8 cylinder
hp/kW	549/404
Construction period	from 2002
Number of produced cars	–

Saleen S7 Turbo

According to a statement by the manufacturer, the Saleen S7, a completely autonomously developed high-performance sports car with a tubular space frame and carbon fibre body, should soon enrich the European market and threaten Ferrari and Porsche. The just one metre-high but nearly two metre-wide racer reaches easily a maximum speed of 350 km/h – while the 100-km/h mark is reached after 3.6 seconds.

Model	Saleen S7 Turbo
Cubic capacity/cylinder	7000 cc/V8 cylinder
hp/kW	760/559
Construction period	from 2005
Number of produced cars	–

Saturn Sky Roadster

The Saturn Corporation (Michigan, USA) was founded already in 1985 as a subsidiary of the General Motors Group. Up to now, Saturn served the market with rather respectable middle class sedans. However, this is expected to change soon. The two-seater roadster "Sky" – with a construction identical to the new Opel GT – should appeal to customers who are in search of a sporty vehicle.

Model	Saturn Sky Roadster
Cubic capacity/cylinder	2384 cc/4 cylinder
hp/kW	180/132
Construction period	from 2006
Number of produced cars	–

SSC Aero SC/T8

As opposed to the prevailing notion that the Shelby Super Cars company is related to Carroll Shelbys company "Shelby North America", the American small series manufacturer SSC is actually totally unrelated to its namesake. The Aero SC/T8 built by SSC costs USD 240.000 – customers who would like to have a back spoiler may obtain one for USD 4000 more. However, in return the air-conditioning, window lifter etc. are serially included.

Model	SSC Aero SC/T8
Cubic capacity/cylinder	6187 cc/8 cylinder
hp/kW	782/575
Construction period	from 2006
Number of produced cars	

SSC Aero SC/T8

After only 3.1 seconds, the SC/T8 is supposed to reach the 100-km/h mark! However, the maximum speed of the only 109 centimetres high and 208 centimetres wide sports care is 400 km/h. In contrast to super sports cars with "streamlined" interiors the SC/T8 pampers its drivers with an audio DVD system (including 10 loudspeakers). The parking becomes a pushover with the help of a reverse camera – according to the manufacturer.

Model	SSC Aero SC/T8
Cubic capacity/cylinder	6187 cc/8 cylinder
hp/kW	782/575
Construction period	from 2006
Number of produced cars	–

Superformance Daytona Coupé

Modelled on the legendary AC Cobra of which in the 1960s also racing coupés have been built, the American company Superformance created a kind of classical reproduction car. The coupé, which is powered by a Ford V8 engine, has a top speed of 310 km/h and features the latest in vehicle technology (tubular space frame, carbon body).

Model	Superformance Daytona Coupé
Cubic capacity/cylinder	6590 cc/8 cylinder
hp/kW	510/375
Construction period	from 2005
Number of produced cars	–

Ultima GTR

The Ultima GTR, which is also distributed in Europe via the Techniccars company, is a street-legal super sports car that was not conceived with luxury and comfort in mind. Instead, it is primarily distinguished by its road performance and driving pleasure. This is why, with an efficiency weight of 2 kg an acceleration time of only 3.2 seconds is required for the car to go from 0 to 100 km/h!

Model	Ultima GTR
Cubic capacity/cylinder	4200 cc/8 cylinder
hp/kW	364/268
Construction period	from 2003
Number of produced cars	–

Ultima CanAm

The Ultima CanAm, the open variation of the GTR, is also purely dominated by technology. The whole interior focuses on functionality, which is why the gearshift lever of the six-step gear (sequential shifting by request) rises casually from the ground. Anatomically designed bucket seats of carbon as well as four-point seat belts provide a real racing sensation.

Model	Ultima CanAm
Cubic capacity/cylinder	4200 cc/8 cylinder
hp/kW	364/268
Construction period	from 2003
Number of produced cars	–

Holden Monaro

Initially, the Holden Monaro, which has been, produced since 1968 was, so to speak, the Australian interpretation of the Chevrolet Camaro. When Holden Automobile later on increasingly adapted the European Opel design, the car became approximately equivalent to the Opel Commodore Coupé. The 480-centimetre long sports coupé of the current series is the only model by Holden that is also available in Europe (England).

Model	Holden Monaro
Cubic capacity/cylinder	5665 cc/8 cylinder
hp/kW	354/260
Construction period	2001 – 2006
Number of produced cars	–

Holden Monaro HSV

In contrast to the standard design version of the Monaro, the Monaro HSV has actually a little more bite. In addition, this version which requires only 5.3 seconds to accelerate from 0 to 100 km/h is equipped with a six-step gear – while its maximum speed is about 242 km/h. In addition to the English market, the HSV is also available in the USA where it is offered, however, as a Pontiac GTO.

Model	Holden Monaro HSV
Cubic capacity/cylinder	5665 cc/8 cylinder
hp/kW	408/300
Construction period	2001 – 2006
Number of produced cars	–

Holden Efijy

The Australian brand Holden, which is integrated into the General Motors Group created with the Efijy – in contrast to other manufacturers – not a show car with modern and futuristic lines, but an eye-catcher designed in the retro style. Nevertheless, it is no longer the 520 centimetre long carriage from yesteryears, because under its rounded exterior the achievement-optimised technology of the current Corvette is hidden.

Model	Holden Efijy
Cubic capacity/cylinder	4600 cc/8 cylinder
hp/kW	655/482
Construction period	2005
Number of produced cars	unique piece

Holden Efijy

The design of the Efijy is certainly modelled on the shape of the Holden FJ built in 1953. However, the FJ still did not feature comfortable details like a touch-sensitive display with a touch screen that can be used, for example to adjust the air-conditioning of the Efijy. Other technical highlights like keyless access and starting system round off the equipment of the show car.

Model	Holden Efijy
Cubic capacity/cylinder	4600 cc/8 cylinder
hp/kW	655/482
Construction period	2005
Number of produced cars	unique piece

Hulme F1 Champion 1967

The New Zealand native Tony Parker has been working for many years on a super sports car similar to the ones that participate in the Formula 1. Parker' sports car which can be driven on the street without any problems is put on the market under the name " Hulme F1 champion 1967". As a reminder – Denis Hulme was the Formula 1 world champion of 1967. The engine technology of the Hulme originates, incidentally, from BMW.

Model	Hulme F1 Champion 1967
Cubic capacity/cylinder	4995 cc/8 cylinder
hp/kW	450/331
Construction period	from 2007
Number of produced cars	–

Joss

If everything works out, the Joss built in Australia will soon become available. The super sports car is intended to threaten its Italian competitors. Due to an ingenious underbody construction, the racer developed by Matt Thomas can do without annoying spoilers. The engine output is, incidentally, transmitted to the rear wheels by means of a Porsche six-speed gear.

Model	Joss
Cubic capacity/cylinder	6800 cc/8 cylinder
hp/kW	500/368
Construction period	from 2006
Number of produced cars	–

Honda NSX

Although the NSX 2005 remained in the company's product range from 1990 to 2005, Honda only produced 18,000 units of the high-tech car in 15 years. The car that was produced in an all-aluminium construction method was something quite special for individualists. Its cult status will continue unchanged in the future, because vintage car collectors are already considering the NSX with a top speed of 270 km/h to be a collectable classic.

Model	Honda NSX
Cubic capacity/cylinder	2977 cc/6 cylinder
hp/kW	255/188
Construction period	2002 – 2005
Number of produced cars	–

Honda NSX-R

The NSX Type R was especially derived for drivers enthusiastic about racing. The model that is even suited for competitive racing moved on broader tyres and disposed of a great amount of strength. In return, it had to make do without amenities such as air-conditioning and part of its interior equipment for weight reasons. The NSX-R, which is only constructed as a right-hand drive car, cost about DM 100.000 in 2002.

Model	Honda NSX-R
Cubic capacity/cylinder	3179 cc/6 cylinder
hp/kW	280/206
Construction period	2002 – 2004
Number of produced cars	–

Honda Integra Type R

With the currently released sports version of the Integra, Honda serves at the moment only the Japanese market. Apart from a slight face-lift, the car has changed only slightly in terms of appearance from its predecessor. The Type R moves on 17-inch light alloy wheels and is powered by a two-litre iVTEC engine that exerts its power via a six-step gear to the street.

Model	Honda Integra Type R
Cubic capacity/cylinder	1998 cc/4 cylinder
hp/kW	220/162
Construction period	ab 2002
Number of produced cars	–

Honda HSC

Shortly after Honda had proclaimed the end of the production of the NSX in 2005, they went on the offensive and announced under reservation the car's sporty successor for the year 2008. There are no details or photos to see yet, but the specialist press assumes that the succession model could consist of the production ready version of the HSC study.

Model	Honda HSC
Cubic capacity/cylinder	3400 cc/6 cylinder
hp/kW	300/221
Construction period	2003
Number of produced cars	unique piece

Infinity G 35 Coupé

In 1989 Nissan presented its noble brand Infinity at the Motor Show in Detroit. From that point on they planned to market a number of especially high-quality series under this label – only in the American market. The calculated effort paid off, and the model range, which at first only consisted of sedans, was complemented in 2002 with a coupé, the G 35.

Model	Infinity G 35 Coupé
Cubic capacity/cylinder	3498 cc/6 cylinder
hp/kW	298/219
Construction period	2002 – 2006
Number of produced cars	–

Honda HSC

With the HSC (Honda Sport Concept) concept car, Honda received a great amount of attention at the 2004 Tokyo Motor Show. Compared to the discontinued NSX, the HSC has a larger wheel base (266 centimetres), while its body also consists of lightweight construction materials. The interior is designed in a sporty fashion, while behind the seats a V6 engine dominates the output.

Model	Honda HSC
Cubic capacity/cylinder	3400 cc/6 cylinder
hp/kW	300/221
Construction period	2003
Number of produced cars	unique piece

Lexus SC 430

To be able to place a new series of high-quality luxury models on the market in style – especially on the American market – Toyota founded in the 1980s the in-house brand "Lexus". In addition to various sedans, they showed in 1999 the prototype of a convertible with a retractable metal top – while two years later the serial production of the concept car was launched under the name SC 430.

Model	Lexus SC 430
Cubic capacity/cylinder	4293 cc/8 cylinder
hp/kW	286/211
Construction period	2001 – 2005
Number of produced cars	–

Infinity G 35 Coupé

Following a carefully carried out face lift, the big G 35 Coupé entered its second round in spring 2006. The car with a top speed of 250 km/h practically consists of the US version of the model Nissan 350 Z Coupé. Similar to this model, the G 35 is likewise a rear-wheel driven 2 + 2-seater. The measurements of the G 35 are [illegible] length x 181 centimetres width x 139 centimetres height.

Model	Infinity G 35 Coupé
Cubic capacity/cylinder	3498 cc/6 cylinder
hp/kW	298/219
Construction period	from 2006
Number of produced cars	–

Lexus SC 430

It takes only 25 seconds to transform the closed SC 430 into a convertible. Actually, the special feature of this car is its ingenious metal roof, which can be stacked into the boot controlled by electric motors. Because this folding construction takes up a relatively large amount of storage space, the storage volume of the boot is greatly reduced when driving with the top down.

Model	Lexus SC 430
Cubic capacity/cylinder	4293 cc/8 cylinder
hp/kW	304/224
Construction period	from 2005
Number of produced cars	–

Lexus SC 430

For safety reasons, Lexus has electronically limited the maximum speed of the SC 430 to 250 km/h – theoretically the velvety-soft running V8 engine could enable the 453 centimetres long convertible coupé to reach the 300-km/h mark. Since autumn 2005 – following very carefully carried out changes in details – Lexus is producing the car already in its second edition.

Model	Lexus SC 430
Cubic capacity/cylinder	4293 cc/8 cylinder
hp/kW	304/224
Construction period	from 2005
Number of produced cars	–

Lexus LF-A

A strikingly prominent wedge-shape, a strong V8 front engine coupled with perfect processing and a luxurious interior – this is what the entry of Lexus into the league of the super sports cars could look like. The concept car LF-A that was revealed on the occasion of the Detroit Motor Show indicates how Lexus intends to position itself in the league of high-performance sports cars.

Model	Lexus LF-A
Cubic capacity/cylinder	4995 cc/8 cylinder
hp/kW	500/368
Construction period	2005
Number of produced cars	unique piece

Lexus LF-A

From a cubic capacity of just five litres, the LF-A draws enough strength to join the the club of the super sports cars with a top speed in excess of 300 km/h. Thanks to ingenious aerodynamics it will be able to even reach 320 km/h – the propulsion technology and power transmission are based on the technical developments of the Formula 1.

Model	Lexus LF-A
Cubic capacity/cylinder	4995 cc/8 cylinder
hp/kW	500/368
Construction period	2005
Number of produced cars	unique piece

Mazda Cosmo 21

In 1967, Mazda released with the Cosmo Sport the first model that was powered by a Wankel engine. Today, the manufacturer is the global leader in this area and continues to be very innovative. In 2002 , the Cosmo 21 was exhibited at the Tokyo Motor Show. In terms of appearance this concept car is reminiscent of the Cosmo of 1967. It is based on the MX-5 and equipped with the engine of the RX-8.

Model	Mazda Cosmo 21
Cubic capacity/cylinder	2 x 654 cc
hp/kW	231/170
Construction period	2001
Number of produced cars	unique piece

Mazda RX-8

The RX-8 is a quite special car, because it was the first four door sports car in automotive history that was equipped with a planetary piston engine. Of the four doors, the two small ones in the rear open in the direction opposite to that of traffic, and by doing away with the B column, the car offers a degree of comfort when getting into it that one would not expect from a sports car.

Model	Mazda RX-8
Cubic capacity/cylinder	2 x 654 cc
hp/kW	192/141
Construction period	from 2002
Number of produced cars	

Mazda RX-8

At the core of the RX-8 is a planetary piston engine that is the most powerful of its construction style without a turbocharger. Because of the rotation of two of the triangular-shaped rotors, a perfect balance of the mass is achieved, and the engine runs smoothly like a six-cylinder piston engine. However, including all the additional engines it weighs a mere 140 kilograms and is barely bigger than a four-cylinder serial engine!

Model	Mazda RX-8
Cubic capacity/cylinder	2 x 654 cc
hp/kW	231/170
Construction period	from 2003
Number of produced cars	–

Mazda MX-5

In the third edition also, the design of the MX-5 was only slightly changed. The little sports car had already bid its pop-up headlights farewell in the second edition, yet the front section had to undergo a face-lift once again. The slightly expanded measurements of the car in combination with prominently designed wheel arches now provide it with a rather grand appearance.

Model	Mazda MX-5
Cubic capacity/cylinder	1798 cc/4 cylinder
hp/kW	126/93
Construction period	from 2006
Number of produced cars	–

Mitsubishi Eclipse Concept-E

At the Motor show in Detroit, Mitsubishi's exhibit included the concept car Eclipse Concept-E. The special feature of this agile unique piece is its hybrid-electric power train. With this concept and its excellent driving dynamics coupled with low pollutant emissions and minimum fuel consumption, Mitsubishi was able to implement a well-calibrated balance between performance and environmental protection.

Model	Mitsubishi Eclipse Concept-E
Cubic capacity/cylinder	n.s.
hp/kW	n.s.
Construction period	2004
Number of produced cars	unique piece

Mitsubishi Lancer Evo IX

In the end of 2005, Mitsubishi launched the by now ninth generation of the Lancer Evo model. Thanks to an overhauled engine, the sporty powerhouse reaches a maximum speed of 250 km/h – while the acceleration to the 100-km/h mark is achieved in 5.7 seconds. In line with its exterior, the Evo IX is serially equipped with Recaro sports seats, aluminium sport pedals and a switchboard made of Titan.

Model	Mitsubishi Lancer Evo IX
Cubic capacity/cylinder	1997 cc/4 cylinder
hp/kW	280/206
Construction period	from 2005
Number of produced cars	–

Mitsubishi Eclipse Spyder

The Eclipse Spyder, shown for the first time at the North American International Auto Show of 2006, was definitely the star of its model class. Its success was rather expected – since this model was developed by the American Mitsubishi design centre (in Cypress/California). The basic concept of the Spyder is actually derived from the Eclipse coupé produced since mid-2005.

Model	Mitsubishi Eclipse Spyder
Cubic capacity/cylinder	2378 cc/4 cylinder
hp/kW	162/119
Construction period	from 2006
Number of produced cars	–

Mitsubishi Eclipse Spyder

The Eclipse Spyder is equipped with a relatively simply designed soft top. Because it is operated electrically, the car can be transformed comfortably and quickly into a breezy convertible. In addition to the 2.4-litre engine, Mitsubishi also offers a six-cylinder version of this car. This stronger version disposes alternatively of a manual six-step gear or a five step automatic transmission.

Model	Mitsubishi Eclipse Spyder
Cubic capacity/cylinder	3828 cc/6 cylinder
hp/kW	263/196
Construction period	from 2006
Number of produced cars	–

Nissan 350 Z Coupé

The first concept car of the 350 Z was shown in 1999 at the Detroit Motor Show. While it received the desired attention, it was rejected again however, because it was considered to be too heavily into the retro look. With support of the Nissan president Carlos Ghosn, the Nissan design studios in Japan, Germany and the USA subsequently competed to create the final design.

Model	Nissan 350 Z Coupé
Cubic capacity/cylinder	3498 cc/6 cylinder
hp/kW	280/206
Construction period	from 2002
Number of produced cars	–

Nissan 350 Z Coupé

The fifth generation of the legendary Z sports car series was available for purchase in autumn 2002, initially only in the USA. In Europe, the coupé was introduced in late 2003 and complemented in 2005 with the roadster version. Globally, the new Z-series has already received more than 40 awards and prizes, substantially strengthening the position of Nissan within the competitive sports car segment.

Model	Nissan 350 Z Coupé
Cubic capacity/cylinder	3498 cc/6 cylinder
hp/kW	280/206
Construction period	from 2002
Number of produced cars	–

Nissan 350 Z Roadster

Similar to the coupé, the 350 Z Roadster also embodies the mindset, style and tradition of the famous Z-series by Nissan. Even if concrete work on it began only after the release of the coupé, the roadster version of the Z did not come unexpectedly for the Nissan designers. After all, it had been included from the start in the production plan.

Model	Nissan 350 Z Roadster
Cubic capacity/cylinder	3498 cc/6 cylinder
hp/kW	280/206
Construction period	from 2005
Number of produced cars	–

Nissan 350 Z Roadster

The most notable feature of the Roadster is its soft top, which only takes 20 seconds to raise or lower. When lowered, it completely stows away under a tonneau panel made from plastic, painted the same colour as the car. When the "bonnet" is down, one can see a pair of sculpted domes that are aerodynamically optimised. They serve to direct the air towards the rear reducing the amount of turbulent air entering the passenger cabin.

Model	Nissan 350 Z Roadster
Cubic capacity/cylinder	3498 cc/6 cylinder
hp/kW	280/206
Construction period	from 2005
Number of produced cars	–

Subaru Impreza WRX STi

With the Impreza, Subaru demonstrates that it is possible to create a sports device from a sedan. The core of the four-door model is a Boxer engine. The 2.5-litre engine is available in three output levels – 230 hp, 280 hp and 320 hp – of which the latter version is supported by a turbo charger. Depending on to the power output of the engine, the maximum speed of the WRX lies between 230 km/h and 255 km/h.

Model	Subaru Impreza WRX STi
Cubic capacity/cylinder	2457 cc/4 cylinder
hp/kW	230/169
Construction period	from 2002
Number of produced cars	–

Toyota Celica TS

With the Celica TS, introduced to the market in the model year 2001, fans of this series could acquire the Celica model with the highest horsepower of all times. Thanks to its lightweight construction the absolutely newly developed car lost quite an amount of weight compared to its predecessor. Of course, this benefited the maximum speed, which now reached 225 km/h.

Model	Toyota Celica TS
Cubic capacity/cylinder	1794 cc/4 cylinder
hp/kW	192/141
Construction period	2000–2006
Number of produced cars	–

Toyota MR 2 Roadster

For the start of the third MR-2 generation Toyota had submitted the car to a thorough rejuvenation process. The expectation that this would once again boost the sales figures was fulfilled. The new MR 2 offered from 2000 offers a number of surprises – the targa had been transformed into a roadster that was now based on an absolutely newly designed platform.

Model	Toyota MR 2 Roadster
Cubic capacity/cylinder	1794 cc/4 cylinder
hp/kW	140/103
Construction period	from 2000
Number of produced cars	–

Toyota MR 2

Thanks to its construction as a Roadster, the MR 2 with a top speed of 210 km/h can be driven in three ways. It can either be driven completely open or with a closed top. In addition, it can also be equipped with a hard top (at a surcharge) and thus be used without limitations in disagreeable weather.

Model	Toyota MR 2
Cubic capacity/cylinder	1794 cc/4 cylinder
hp/kW	140/103
Construction period	from 2000
Number of produced cars	–

Toyota MR 2 Competition

For the start of the 2002 racing season, Toyota introduced an exclusive and limited sports edition of the MR 2 – the MR 2 Competition. Based on the serial model, the sports model, which was available exclusively painted "volcano red", is distinguished by various white decorative applications on the bonnet and along the sides.

Model	Toyota MR 2 Competition
Cubic capacity/cylinder	1794 cc/4 cylinder
hp/kW	140/103
Construction period	2002
Number of produced cars	100

Toyota Alessandro Volta

The Italian coach maker Italdesign has created a sports car of superlatives – the Toyota Alessandro Volta. According to Italdesign, this is what the sports car of the future will look like – it is extremely environmentally friendly, because it is powered in addition to an internal combustion engine also by two added electric motors!

Model	Toyota Alessandro Volta
Cubic capacity/cylinder	3300 cc/6 cylinder
hp/kW	408/300
Construction period	2003
Number of produced cars	unique piece

Toyota Alessandro Volta

In spite of a length of 430 centimetres only three persons can be seated in the Toyota Alessandro Volta – and that only side by side! Its modern hybrid technology takes up remaining interior space. This technological combination provides 408 hp – which make the low-riding car reach 250 km/h and accelerate from 0 to 100 km/h in 4.06 seconds.

Model	Toyota Alessandro Volta
Cubic capacity/cylinder	3300 cc/6 cylinder
hp/kW	408/300
Construction period	2003
Number of produced cars	unique piece

DC Go

Dilip Chhabria, founder of the DC design studio, was already frequently represented at European automotive shows. In 2006, the Indian native presented the DC Go, a stylistically very successful sports car study. Based on the British Noble M 12, the unique piece features a luxurious leather interior; under the bonnet it is powered by a serial Ford engine.

Model	DC Go
Cubic capacity/cylinder	2500 cc/[illegible] cylinder
hp/kW	310/228
Construction period	2005
Number of produced cars	unique piece

Hyundai Coupé 2.7 V6 GLS

Although the Hyundai brand is a synonym for especially inexpensive cars, the Korean manufacturer is also trying to gain a foothold in the sports car class. According to factory specifications, the fluent lines of the GLS are exactly in line with the taste of its customers – a principal reason why the model with a top speed of 220 km/h and a six-cylinder engine sells very well on the European market.

Model	Hyundai Coupé 2.7 V6 GLS
Cubic capacity/cylinder	2657 cc/6 cylinder
hp/kW	167/123
Construction period	from 2005
Number of produced cars	–

Laraki Borac

Up to now, Morocco has played no role in the more than 100-year-old automotive history. This could soon change, because at the Laraki company in Casablanca the plan of a super sports car is close to its completion. Laraki, a manufacturer of exclusive deluxe yachts, would thus also offer an attractive locomotion means for the street.

Model	Laraki Borac
Cubic capacity/cylinder	5495 cc/8 cylinder
hp/kW	600/441
Construction period	from 2007
Number of produced cars	–

Laraki Borac

The Borac was developed under the direction of the company founder Abdeslam Laraki. Laraki, who had been trained by the Swiss tuner Sbarro developed the aluminium chassis himself utilising Lamborghini components. It is powered by a Mercedes-Benz engine that allows it to reach a speed of 310 km/h.

Model	Laraki Borac
Cubic capacity/cylinder	5495 cc/8 cylinder
hp/kW	600/441
Construction period	from 2007
Number of produced cars	–

Laraki Fulgura

Alternatively to the Borac, Laraki also plans the production of another sports car, the Fulgura. It was already presented in the prototype stage at several European automotive fairs. The low-riding super sports car is planned to be equipped with a V8 engine placed before the rear axle and expected to accelerate in 3.8 seconds to 100 km/h.

Model	Laraki Fulgura
Cubic capacity/cylinder	5495 cc/8 cylinder
hp/kW	600/441
Construction period	from 2007
Number of produced cars	–

Laraki Fulgura

Apparently the Laraki Fulgura will be serially equipped with a V8 engine. Yet nevertheless there also exists a prototype with a V12 engine. This version is also described in the Laraki company brochure. Thanks to its light carbon fibre bodywork this model reaches a maximum speed of 350 km/h according to the manufacturer's information.

Model	Laraki Fulgura
Cubic capacity/cylinder	5995 cc/12 cylinder
hp/kW	680/500
Construction period	2002
Number of produced cars	unique piece

Cobra Arbitrage GT

Cobra International – according to the company's own information the world's largest manufacturer of surfboards – wants to establish itself from 2008 also in the sports car business. Strictly speaking, the company headquartered in Thailand will actually implement an automobile design developed by the American designer Kevin Gallahan and the Canadian tuner Alan Mertens.

Model	Cobra Arbitrage GT
Cubic capacity/cylinder	3200 cc/6 cylinder
hp/kW	250/184
Construction period	from 2008
Number of produced cars	max. 350

Cobra Arbitrage GT

If the Cobra Arbitrage is introduced in a limited series as planned in January 2008, customers can choose between two versions. In its standard version, the two-seater is equipped with a tuned V6 engine from the VW Golf R32 (250 hp) while alternatively there is also supposed to be a Twin Turbo variation available with 550 hp. The price of the racer will probably be in the EUR 100,000 range.

Model	Cobra Arbitrage GT
Cubic capacity/cylinder	3200 cc/6 cylinder
hp/kW	550/404
Construction period	from 2008
Number of produced cars	max. 350

Russo-Baltique Impression

Russo-Baltique, an automobile manufacturer absolutely unknown in Europe, plans to build cars again after a break of several decades. However, these will be limited to only 15 units that will be priced accordingly – the Impression is expected to cost USD 1.0 million! The Russian deluxe vehicle (including its engine) is based on a Mercedes CL 65 AMG – while its baroque bodywork is made of carbon fibre.

Model	Russo-Baltique Impression
Cubic capacity/cylinder	6000 cc/12 cylinder
hp/kW	[illegible]/408
Construction period	from 2007
Number of produced cars	–

Index

A

B

C

D

F

G

H

I

J

K

L

M

N

O

P

Q

R

S

T

U

V

W

Y

Z

OS-MS 8